REFRACTORY LININGS

MECHANICAL ENGINEERING

A Series of Textbooks and Reference Books

Editor

L. L. Faulkner

*Columbus Division, Battelle Memorial Institute
and Department of Mechanical Engineering
The Ohio State University
Columbus, Ohio*

Additional Volumes in Preparation

Mechanical Engineering Software

REFRACTORY LININGS

THERMOMECHANICAL DESIGN AND APPLICATIONS

CHARLES A. SCHACHT
Schacht Consulting Services
Pittsburgh, Pennsylvania

CRC Press
Taylor & Francis Group
Boca Raton London New York

CRC Press is an imprint of the
Taylor & Francis Group, an **informa** business

CRC Press
Taylor & Francis Group
6000 Broken Sound Parkway NW, Suite 300
Boca Raton, FL 33487-2742

First issued in paperback 2019

© 1995 by Taylor & Francis Group, LLC
CRC Press is an imprint of Taylor & Francis Group, an Informa business

No claim to original U.S. Government works

ISBN-13: 978-0-8247-9560-3 (hbk)
ISBN-13: 978-0-367-40190-0 (pbk)

Library of Congress Cataloging-in-Publication Data

Schacht, Charles A.
    Refractory linings: thermomechanical design and applications / Charles A. Schacht.
        p. cm. — (Mechanical engineering; 95)
    Includes bibliographical references and index.
    ISBN 0-8247-9560-1 (acid-free paper)
    1. Refractory materials—Thermomechanical properties.   I. Title.
II. Series: Mechanical engineering (Marcel Dekker, Inc.); 95.
TA418.26.S32   1995
666'.72—dc20   94-24950
                                                                         CIP

**Visit the Taylor & Francis Web site at
http://www.taylorandfrancis.com**

**and the CRC Press Web site at
http://www.crcpress.com**

In memory of my wife, Sue Schacht,
and my father, Lawrence Schacht

Preface

Refractories have always been a necessary part of industry. High-temperature industrial processes require refractory materials to insulate the process, assist in controlling the process temperatures and insulate the process heat from the support structure. Typically, heat transfer characteristics (thermal material properties) for the various kinds of refractories are well defined by refractory manufacturers because of the immediate need for this information in evaluating refractory lining systems. Refractories deteriorate and wear due to the chemical and mechanical conditions imposed on them. Mechanical and chemical deterioration can occur either in concert or separately. In some industrial processes the refractories last only ten days, while in others the refractory can last years or decades. Because of the increasing competitiveness in industry, various means of extending refractory life are being pursued, and investigations regarding the structural/mechanical behavior of refractory lining systems are gaining considerable interest.

Refractory lining systems behave in a very complex manner mechanically. If mechanical wear is to be investigated and better lining designs pursued, the structural/mechanical behavior of refractory lining systems must be understood. Since limited information is available in this area, *Refractory Linings: Thermomechanical Design and Applications* concentrates on all structural and mechanical aspects of refractory lining systems. This book is not intended to be a treatise on structural mechanics, but rather a resource for understanding the fundamentals of the structural/mechanical behavior of refractory lining systems and the associated deterioration and wear of refractory lining systems.

The basics of heat transfer are also presented, since heat transfer is a vital part of the investigation of refractory lining systems. Mechanical investigations of refractory linings are, therefore, more appropriately described as thermomechanical investigations. That is, the heat transfer investigation and the mechanical investigation are coupled parts of a typical lining investigation.

Chapter 1 provides a general introduction to the thermomechanical investigation of refractory lining systems.

The types of loadings imposed on refractory lining systems are discussed in Chapter 2. *Stress-controlled* and *strain-controlled* loads are defined. The recognition of the load type(s) to which refractory linings are exposed is the most important basis for understanding the refractory lining investigation.

Chapter 3 introduces the initial fundamental assumptions for the mechanical material properties of refractories. The basic concepts of material static and creep behavior are introduced, along with the initial assumptions regarding the mechanical behavior of refractory lining materials.

A considerable number of ASTM tests are used to evaluate the mechanical properties of refractories. Chapter 4 addresses the limitations of the ASTM strength tests.

Chapter 5 presents a rational approach for selecting the strongest or best refractory. Often the crushing strength is not the appropriate material property for choosing the strongest refractory. Selection of the best refractory is dependent on the predominant load type imposed on the refractory material.

Chapters 6, 7, and 11 present material property data, including static compressive stress-strain, creep and thermal expansion data, for various typical refractory materials. All are necessary in evaluating the structural/mechanical behavior of a refractory lining system.

Chapter 8 introduces the basic aspects of the mechanical behavior of mortared and dry joints. The behavior of the joint should be considered when expansion allowance concerns are addressed.

In Chapter 9, the concept of the *hinge* in refractory lining systems is introduced. A hinge occurs at a brick joint in which only a small portion of the joint length is subjected to the compressive load.

The effects of the stress-strain state on the strength of refractory material are described in Chapter 10. As with most brittle materials, the dimensionality of the stress-strain state has a profound influence on the compressive strength and much less influence on the tensile strength.

Chapters 12 through 15 introduce the basic concepts of refractory lining behavior for four types of lining geometries: the refractory arch, the cylindrical lined shell, the spherical refractory dome and the square and rectangular refractory-lined vessels.

Chapter 16 is devoted to analytical investigations on the mechanical deterioration of refractory linings as related to tensile fracture. Typically, previous investigators dealing with transient thermal tensile stresses, such as thermal shock and other similar transient effects, have overlooked the effects of the vessel shell restraint contributing to thermal tensile fracture, even during steady-state thermal conditions. Quite often the effects of vessel restraint can completely alter the transient thermal stresses during both heatup and cooldown. Material design objectives are also presented with regard to work-of-fracture (WOF) testing and the desirable WOF curves based on load type(s) imposed on the refractory material.

Finally, Chapter 17 describes the results of three separate studies on three different refractory lining systems. All three studies compare the analytically predicted results with field measurements and field observations. The three lining systems are a steelmaking ladle, a refractory sprung arch and a spherical refractory lining.

The author would like to acknowledge the following organizations for use of data on refractory materials: Harbison-Walker Refractories, Division of Indresco, Inc., Pittsburgh, PA; National Refractories & Minerals Corp., Livermore, CA; J. E. Baker Co., York, PA; North American Refractories Co., Technical Center, State College, PA; and Ceram Research, Penkhull, England. Special thanks also goes to Carolyn Brown, Desktop Publishing Services, for her patience and understanding in interpreting and processing the original handwritten text.

Charles A. Schacht

Contents

1
Introduction and General Overview

I. INTRODUCTION

This book is intended to be multifunctional, being able to satisfy the needs of the user, installer, designer and investigator of refractory lining systems. For the plant engineer needing some insight into refractory problems on lining spalling and other forms of mechanical deterioration, the example studies should be of value. For the installer, the importance of installation procedures should become more evident from the various lining geometry studies. To the designer and investigator of linings needing to understand the thermomechanical behavior of lining systems, this book should be of great assistance. Finally, for those wanting more material property data on various kinds of refractories, this book can be used as a reference or guide to selecting refractories for specific applications. Therefore, the appropriate chapters can be selected based on the user's requirements.

Refractories are used in industrial processes that require high operating temperatures. Refractories serve several important roles: to control process temperatures, to insulate the temperature-sensitive supporting structure and to conserve heating energy by minimizing heat loss. Typical manufacturing industries requiring refractories include the iron and steel industry, and the non-ferrous (aluminum, zinc, lead, magnesium, copper, etc.) industries. Refractories are also vital in the production of glass, cements, and other materials produced by basic industries. Refractories are also a necessary part of the fossil fuel power industry. Here they are used to line the boiler furnaces, exhaust stacks and other power plant process structures. Refractories have been of particular importance in the coal gasification research [1] in which refractory linings have been given special attention using state-of-the-art technology in the design of the lining. The environmental industry employs the use of refractories in the various types of furnaces used to break down temperature-degradable waste materials. The petroleum industry requires refractories to insulate vessels used in the fractional distillation of crude oil. There are many other industries too numerous to mention here that also use refractories to contain high-temperature processes.

This book does not address the chemistry side of refractories. There is considerable information in the literature [for example, references 2-4] which deals quite adequately with refractory chemistry as related to compatibility with the process environment and as related to chemical deterioration and wear of refractories. Considerably less is understood and has been written regarding the mechanical related behavior, wear and deterioration of refractories. There are numerous examples [5-8] of indirect approaches to refractory wear in terms of laboratory inferential testing of refractories, use of limited mathematical methods or investigative work on lining wear in full-scale production vessels. These types of investigations provide insight into lining wear, but do not provide an understanding regarding the actual detailed or totally integrated stress/strain behavior within the refractory lining as related to the lining fracture, wear and deterioration. Most refractory lining systems are constructed of refractory brick shapes or block although monolithic castable refractories are gaining considerable attention as another reliable alternative in the design of refractory lining systems. The objective of this book is to provide an understanding with regard to the mechanical fracture, wear and deterioration of refractories along with an understanding of the mechanical or structural behavior of the complete refractory lining system.

II. THE REFRACTORY STRUCTURE

Refractory linings constructed of brick shapes can be classified as a block type structure in which various refractory brick shapes [9-11] are used to construct a refractory lining. Figure 1.1 [10] describes the brick shapes used to construct an arch roof. Often block type structures are made up of a wide range and

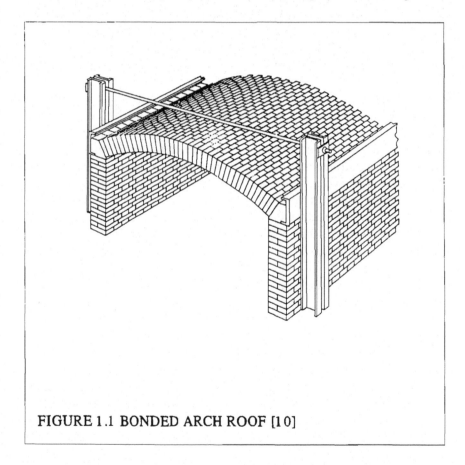

FIGURE 1.1 BONDED ARCH ROOF [10]

variety of shapes and sizes of block to arrive at a desired geometry. Dome linings and cylindrical lined vessels and other lining geometries are similarly constructed of appropriately selected shapes provided by the refractory manufacturer.

Many types of block type or stone arch structures were constructed during the Roman Empire [12]. During the latter centuries of Europe, more sophisticated stone block structures evolved. The Romans used stone arches to support the aqueducts and a combination of arches and domes to construct buildings. During the Gothic period, the French developed, painfully by trial and error, the elegant flying buttresses in conjunction with the massive cathedrals. During these early times throughout England [13], the stone arch was used to construct bridges with significant spans. Investigative work was continued [14-16] in England to better understand the behavior of stone arches. There are still ongoing investigative activities in England pursuing the understanding of stone arches.

Stone was used in these early times because of its availability and compressive strength and because stone could be chiseled and shaped into a desirable block geometry. Because mathematics, material technology and other related sciences were in their infancy, suitable block structures could only be developed by trial and error. The shape, size and other aspects of the individual blocks along the overall geometry of a new structure were not always adequate for the loads and other service environments imposed on the structure. Records show that some of the early structures collapsed, and quite often catastrophically. The key to success in all of these structures was the adequacy of the stone block size and corresponding structure geometry such that they did not crush or fracture under compressive load. The evolved shape of the overall structure resulted in minimum shear across joints and insignificant tensile loading normal to block joints while maximizing compressive loads normal to block joints. The dry joints between the stone blocks cannot sustain tensile loads. Even mortared joints have limited strength and can only support exceedingly small tensile loads. Despite these restrictions, the block structure evolved into successful support structures. Not only were these structures functional, but they also have a grace and beauty that is still admired today.

With the advent of ironmaking, the development of metal structural components resulted in a change in the basic design philosophy of block type or stone structures. That is, metal structural components could sustain both tensile and compressive loads and were superior to stone block structural components (the stone block joints) that were limited to only compressive loads. Stone structures evolved into a combination of stone and metal components. Because of the strength and versatility of metal, the use of these metal components replaced the block stone components. The flying buttress was no longer needed to support the cathedral lateral roof forces. Later, as science and technology progressed, sophisticated structural analytical methods were developed allowing engineers to predict mathematically the behavior of structural systems and to design more reliable structures. As a result, stone block structures had become obsolete before the comprehensive structural analytical methods were developed that could explain the complex behavior of these structures. Contemporary structures constructed of homogeneous strength-bearing materials (steel and reinforced concrete) that support both tensile and compressive stress states have received and continue to receive considerable attention in all of the fields of structural research with regard to safety, optimum strength, minimized material usage and other structurally related aspects.

Refractory structures exhibit similar structural behavior as the block type or stone structures. However, refractory structures are exposed to a more complicated load environment: thermal expansion loading. The refractory lining is typically exposed to high temperatures resulting in complicated stress/strain environments within the lining structure, and the refractory material properties tend to vary as a function of temperatures. High temperatures have a profound influence on the refractory material properties. At high temperatures, refractory materials no longer remain totally elastic. Instantaneous plastic and time-dependent creep straining takes place. As a result, the structural investigation often requires the use of computerized structural analysis methods that can deal simultaneously with these complex material and structural behaviors.

Presented in this book are the basic elements needed to conduct thermal and mechanical structural investigations of the refractory lining structures. Also included are thermal material data and mechanical material data vital to conducting a refractory lining investigation. Since computer methods are a necessary tool for conducting a reliable refractory lining investigation, basic information regarding popular current numerical analytical methods used for structural analysis is also provided. For those less interested in an understanding of the structural behavior, examples of a variety of refractory lining investigations conducted by the author are presented.

III. ANALYTICAL METHODS

Refractory lining systems are typically exposed to a thermal environment that results in expansion of the refractory lining system. As a result, a lining investigation typically consists of two parts. The first is an evaluation of the thermal response of the lining system, which may be either steady-state or transient. Of primary interest in the thermal analysis is the temperature distribution within the lining system. Heat loss and other aspects of the heat transfer analysis may be of interest to some investigators but will not be pursued in this text. The second part of the lining investigation consists of evaluating the thermal expansion, displacements and thermal stresses within the lining system.

One has the choice of various types of analytical methods in solving the thermal and the thermal stress problems. Handbook equations can be used to estimate both the temperature profile in the lining as well as the thermal expansion stresses. However, handbook equations are quite often limited to simplified boundary conditions and simplified material behavior and, therefore, underestimate the complex structural behavior of the most simple refractory lining systems. Handbook equations' results, in most cases, do not match the response of the actual refractory lining system. That is, even for a very simple lining geometry, the complex mechanical behavior of the lining system along with the temperature dependent mechanical behavior of the refractory material make these handbook equations unrealistic in their applicability.

Computerized structural analytical methods have gained considerable popularity in the engineering community especially with the advent of the larger personal computers (PCs) and upgraded structural software that takes advantage of the latest PC capabilities. Most popular is the finite element method (FEM) [17,18]. Basically, the structure to be evaluated is discretized mathematically into a geometric pattern or mesh. The mesh unit is the finite element which has been formulated using algebraic equations to define the unknown variable behavior within the element based on defined quantities of the node points. The finite elements are assembled mathematically to form large matrices which in turn define mathematically the behavior of the total structure. Computers are necessary to solve the large system of algebraic equations thereby defining the unknown variables at each of the nodes. These define the desired structural response in question. Appendix A provides a limited list of currently available FEM programs.

The steady-state thermal problem is defined in matrix form as:

$$[K]\{T\} = \{Q\} \qquad (1.1)$$

where [K] is the thermal conductivity matrix, {T} is the temperature vector (at each node of the mesh model) and {Q} is the heat flow vector (at each node). For transient thermal analysis the matrices become more complex and introduce the effects of time.

The static structural matrices are of the form:

$$[K]\{U\} = \{F\} \qquad (1.2)$$

where {K} is the structural stiffness matrix, matrix {U} is the displacement vector (at each node) and {F} is the force vector (at each node). Additional matrices are formed when inelastic and time effect are included. For those who wish to pursue a more detailed understanding of the finite element method of analysis references 17 and 18 should be reviewed as a starting point.

The assembly of the complete matrices as defined above forms a set of simultaneous equations. The solution of these simultaneous equations results in values for each degree of freedom in the vectors {T} and {Q} for the thermal problem or {U} and {F} in the structural problem.

REFERENCES

1. Materials for Coal Conversion and Utilization, Proceedings Fourth Annual Conference, U.S. Department of Energy, Conf-791014, October 9-11, 1979.

2. Norton, F. H., Refractories, McGraw Hill Book Company, Inc., Fourth Edition, New York, 1968.

3. American Society of Ceramic Engineers, 735 Ceramic Place, Westerville, Ohio 43081-8720.

4. British Ceramic Research Association, Penkhull, Stoke-on-Trent, Great Britain, ST4 7LQ.

5. Brezny, B., Crack Formation on BOF Refractories During Gunning, Iron and Steel Engineers, AISE, Vol. 58, No. 7, pp. 679-682 (1979).

6. Ainsworth, J. H., Calculation of Safe Heat-Up Rates for Steel Plant Furnace Linings, Am. Ceram. Soc. Bull., Vol. 58, No. 7 (1979).

7. Thomson, G. M. and Davies W., Stress and Stain in Furnace Life, Trans. Brit. Ceram. Soc., Vol. 68, Paper No. 36, pp. 269-278 (1969).

8. Kato, I., Morita, Y., Hikami, F., Thermal Stress Formula for Estimation of Spalling Strength of Rectangular Refractories, Tetsu-to-Hagane, Vol. 68, No. 1, pp. 105-112 (1982).

9. Modern Refractory Practices, Third Edition, Harbison-Walker Refractories Co., Pittsburgh, PA, 1950.

10. Harbison-Walker Handbook of Refractory Practice, Second Edition, Harbison-Walker Refractories Co., Pittsburgh, PA, 1980.

11. Refractory Pocket Catalog, A. P. Green, Mexico, MO.

12. Gardner, H., Art Through the Ages, Fourth Edition, Harcourt, Brace and Co., New York, 1959.

13. Snell, G., On the Stability of Arches, Minutes of Proceedings of Institution of Civil Engineers, ICE, London, Vol. 5, No. 1, pp. 439-472 (1846).

14. Pippard, A. J. S., Tranter, E., Chitty, L., The Mechanics of the Voussoir Arch, Journal of Institution of Civil Engineers, ICE, London, Vol. 4, Paper No. 5108, December, pp. 281-306 (1936).

15. Pippard, A. J. S. and Ashby, R. J., An Experimental Study of the Voussoir Arch, Journal of Institution of Civil Engineers, ICE, London, Vol. 10, No. 5177, January, pp. 383-404 (1939).

16. Pippard, A. J. S. and Chitty, L., Repeated Load Tests on a Voussoir Arch, Journal of Institution of Civil Engineers, ICE, London, Vol. 17, Paper No. 5268, November, pp. 79-86 (1941).

17. Przemieniecki, J. S., Theory of Matrix Structural Analysis, McGraw-Hill Book Co., New York, 1968.

18.   Zienkiewicz, Q. C., <u>The Finite Method in Structural and Continuity Mechanics</u>, McGraw-Hill Publishing Co., Limited, London, 1967.

2
Load Types

I. INTRODUCTION

Of most importance in evaluating the structural behavior of refractory lining systems is understanding the concept of load types. The type of load imposed on a structure will have a direct impact on which parameters, such as structural geometry, material properties and stress limits, demand the most attention in optimizing structural response, structural life and other desirable features of the refractory lining structural system.

There are two types of loads [1-3] that can be imposed on a structure. They are defined as *stress-controlled* load and *strain-controlled* load. By understanding the load types imposed on a refractory lining, one can better decide on the most desirable refractory material mechanical properties and refractory lining geometry that will result in the best lining design.

II. STRESS-CONTROLLED LOAD

A stress-controlled load on a refractory structure is an external load, such as a gravity load, pressure load, or other type of mechanical load. By definition, this load is an external load, even when the loading is the structural gravity load. A fundamental structure is defined in Figure 2.1 with an applied stress-controlled load (P).

The primary result of a stress-controlled load is that the stress (S) in the structure can be determined with knowledge of only the structural geometry. In the case of our simplistic structure, the stress is calculated by dividing the load by the cross-sectional area (A) of the structure:

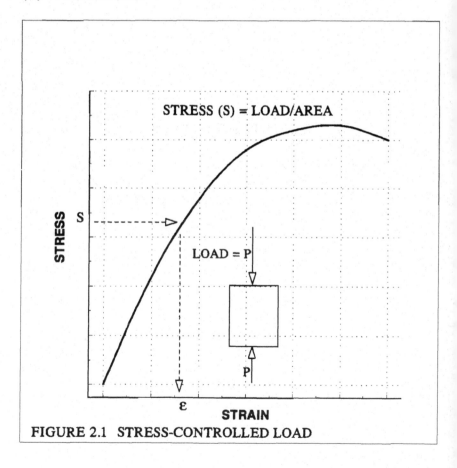

FIGURE 2.1 STRESS-CONTROLLED LOAD

$$S = P/A \qquad\qquad (2.1)$$

Even for more complicated, statistically redundant structures [4,5], only the structure geometry is required to evaluate the stresses within the structure.

The basic requirement of the structure, with respect to a stress-controlled load, is that the structure must satisfy the equilibrium requirements when the structure is subjected to these external loads. The internal stress must equilibrate the external load. Note that mechanical material properties have not entered into the discussion regarding the determination of stresses due to stress-controlled loads. The only need for mechanical material properties (modulus of elasticity, Poisson's ratio, etc.) for stress-controlled load environments is when structural deflections or material strains are to be determined. This concept is illustrated in Figure 2.1; with the known stress, the corresponding strain (ε) can be determined only with knowledge of the material stress-strain curve.

The structure must satisfy the equilibrium requirements for stress-controlled loads. As a result, deformations will not relieve stress-controlled loads. That is, the deflection of the structure will not alter the magnitude of an external load or stress-controlled load. The pressure load, the gravity load or any other stress-controlled loads will remain at the defined value regardless of the structural deformation. The impact of the deformations not altering the stress-controlled load magnitude has a profound influence on how the structure is designed against possible overload. Additional details on the design of structures against possible overload of stress-controlled load will be presented in Section VI.

When the magnitude of the stress-controlled load exceeds the strength of the structural material, unlimited deformations occur. Assuming that the material yields during overload implies that the structure cannot internally equilibrate the external stress-controlled loads. Since the structure cannot support the loading, the structure will collapse.

III. STRAIN-CONTROLLED LOAD

A strain-controlled load on refractory structures is an internal load that is applied to the structure.   More specifically, a strain-controlled load is an imposed strain rather than an imposed stress on the structure.   The strain-controlled load can also be defined as an internal self-equilibrating load.   For the refractory structure, the thermal expansion of the structure is an imposed strain and is classified as a strain-controlled load.   The concept of a strain-controlled load is illustrated in Figure 2.2. The refractory block structure in Figure 2.2 is assumed to be heated from ambient temperature with total temperature change of ΔT.   For the purposes of this simplified example, it is assumed that only vertical expansion occurs on the block structure.   The thermal strain (ε) is

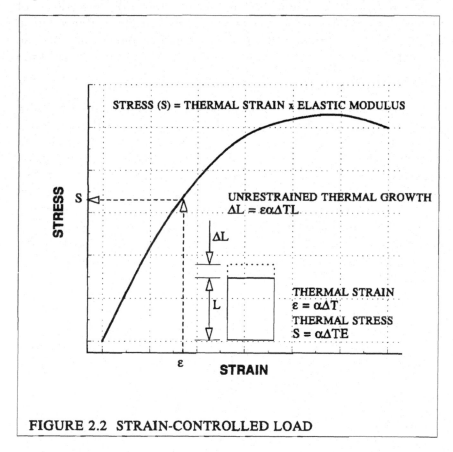

FIGURE 2.2 STRAIN-CONTROLLED LOAD

calculated with the knowledge of one mechanical material property, the coefficient of thermal expansion (α). Therefore, the thermal strain is determined by multiplying the temperature change by the coefficient of thermal expansion:

$$\varepsilon = \alpha \Delta T \qquad (2.2)$$

For a completely unrestrained example block structure, the total unrestrained vertical growth, ΔL, is determined by multiplying the thermal strain by the vertical height, L, of the block:

$$\Delta L = \varepsilon L = \alpha \Delta T L \qquad (2.3)$$

Note here that unlike the stress-controlled loading, the deformation of the structure is finite at the value ΔL. That is, regardless of the material strength, the deformation is limited to the ΔL value.

Refractory linings are typically restrained to some degree by the external support structure. Only through a complete analysis of the total refractory lining and support structure can the interaction and resulting magnitude of restraint between the lining and support be determined. For the purposes of this example problem, it is assumed the refractory block structure is fully restrained. That is, the full magnitude of the thermal growth strain is prevented. Therefore, the heated vertical dimension is identical to the initial ambient dimension. The thermal stress (S), as shown in Figure 2.2, can be calculated by use of the stress-strain relationship for the example refractory material. Assuming a modulus of elasticity of E for the tangent to the point on the stress-strain curve in question, the thermal stress is:

$$S = \varepsilon E = \alpha \Delta T E \qquad (2.4)$$

Other types of strain-controlled loads include shrinkage of the material due to loss of moisture. This type of loading is most likely to occur in castables, gunning and ramming mixes and other materials that are applied in a wet or *green* state. Shrinkage is actually a reverse of the thermal expansion strain. Most expansion strains are positive while most loss of moisture (or shrinkage) strains are negative.

There are other strain-controlled loadings that occur in refractories. In refractory materials growth or swelling of the refractory materials occur due to a cross-coupling of chemistry with temperature and time [3-7]. However, very little information is available with regard to the amount of straining due to these effects or the influence of these effects on the mechanical stress-strain behavior of the refractory material.

In the nuclear industry, neutron bombardment [3,6] results in a swelling or a positive volumetric strain growth of the material. However, these more exotic strain-controlled loadings will not be addressed here.

IV. DESIGN PHILOSOPHY OF STRUCTURES BASED ON LOAD TYPE

With the knowledge of the load types that can be imposed on structures, the design philosophy can now be established. That is, what does one consider as a satisfactory margin of safety against collapse of the structure for a possible overload. The term *factor of safety* is used to define the additional structural adequacy built into the structure to guard against overload and the resulting collapse of the structure.

There are currently several methods used to establish the design of the structure against overload and resulting structural collapse. A portion of these methods are probabilistic and others are more deterministic. That is, some methods deal with the unknowns of defining the loads applied to the structure through the use of probability models. The deterministic methods assume that one can adequately design sufficient safety into a structure by defining limits of stress and strain. A deterministic method will be used here to define the design of the structure against overload, not because of any favoring of deterministic methods, but the objective here is to add clarity to the impact of stress- and strain-controlled loads.

V. SUMMARY

A stress-controlled load is an external mechanical load such as gravity, pressure, or inertia load imposed on a structure. The structure equilibrates the stress-controlled loads. The resulting equilibrating stresses are determined through the knowledge of the structural geometry. With a stress-controlled load, the stress is the known parameter within the structure. The knowledge of the material stress-strain relationship is not required. Only the magnitude of the defined failure stress is required.

A strain-controlled load is an internal self-equilibrating load caused by an imposed strain on the structure such as by thermal straining. Therefore, the strain is known in a structure when subjected to a strain-controlled loading. Knowledge of the material stress-strain relationship is required for analyzing this type loading. Stress resulting from a strain-controlled load can be determined only through the knowledge of the material stress-strain relationship.

With a stress-controlled load, the ability of the structure to sustain the stress-controlled load is based on the value of the ultimate failure stress of the subject structural material. With a strain-controlled load, the ability of the structure to sustain the strain-controlled load is based on the value of the ultimate failure strain of the subject structural material.

REFERENCES

1. Langer, B. F., Design Values for Thermal Stress in Ductile Materials, ASME, Welding Journal Research Supplement, Vol. 37, pp. 411-S., Sept. (1958).

2. Jetter, R. I., Elevated Temperature Design-Development and Implementation of Code Case 1592, ASME Trans., Journal of Pressure Vessel Tech., pp. 222-229, August (1976).

3. Criteria of The ASME Boiler and Pressure Vessel Code for Design by Analysis, in Sections III and VIII, Division 2, ASME (1969).

4. Roark, R. J. and Young, W. C., <u>Formulas for Stress and Strain</u>, Fifth Edition, McGraw-Hill Book Co., New York, 1975.

5. <u>Manual of Steel Construction</u>, Eighth Edition, American Institute of Steel Construction, Inc., 1980.

6. Nuclear Systems Material Handbook, Vol. 1--Design Data, Part 1--Structural Materials, Group 1--High Alloy Steels. (Availability: Hanford Engineering Development Laboratory, Richland, WA).

7. Huang, B. Y. and McGee, T. D., Secondary Expansion of Mullite Refractories Containing Calcined Bauxite and Calcined Clay, Ceramic Bulletin of the Amer. Ceram. Soc., Vol. 67, No. 7, pp. 1235-1238 (1988).

3
Material Properties

I. INTRODUCTION

A variety of tests to define thermal chemical, electrical, thermal and mechanical properties have been developed by The American Society for Testing and Materials (ASTM), the world's largest source of voluntary standards on testing for quantifying the characteristics and performance of materials and, in our case, refractories. Of particular interest here are the tests used to define the mechanical properties of refractories.

As indicated in Chapter 1, the primary interest of this text is the combined thermal and structural or thermomechanical behavior of refractory structures. To this end, the primary interest with respect to material properties are those material properties necessary to conduct thermal and structural investigations of refractory lining systems. To further assist in defining the necessary material

properties for structural investigations, a brief discussion is
provided in this chapter on the limitations (referred to in Section VI
as *inferential* material properties) of some ASTM tests used to
quantify material properties. However, prior to the discussion on
inferential material properties, Sections II through V provide the
necessary background with regard to material properties.

II. MATERIAL PROPERTIES REQUIRED FOR STRUCTURAL ANALYSIS

Refractory lining systems (or refractory structures) are
exposed primarily to thermal expansion forces. Typically, most
refractory linings are restrained against complete free thermal
expansion by the external structural steel. In most cases, the thermal
stresses due to the restraint are considerably greater than the gravity
weight stresses. Therefore, the thermal analysis of the structure is a
necessary part of the refractory lining structural investigation. A
typical lining structural investigation can be classified into two
major parts: the thermal analysis and the mechanical analysis. The
thermal analysis is used to evaluate the lining temperatures while
the mechanical analysis is used to evaluate the resulting thermal
restraint stresses.

Because refractories' primary purpose is to restrict heat flow
of a manufacturing process, refractory manufacturers have
historically provided thermal material properties of their products.
With regard to transient thermal analysis, the required thermal
material properties include the thermal conductivity, specific heat
and density. For steady-state thermal analysis only the thermal
conductivity is required.

The mechanical analysis of the thermal restraint stresses
requires a definition of the mechanical material properties of the
refractories. The most fundamental required properties are the
elastic modulus, Poisson's ratio and the coefficient of thermal
expansion. If the gravity load stresses are to be included in the

structural investigation, then the density of the refractory material is necessary for the mechanical analysis. The refractory lining system is typically exposed to a range of temperatures; therefore, mechanical material properties should be defined for the temperature range of interest. Refractory linings exposed to high process temperatures result in inelastic stress-strain response within the refractory material. At the higher temperatures the mechanical material property requirements have to be expanded to include the inelastic mechanical material properties (plastic and creep).

Structural investigations of refractory lining systems can be classified as either short-term or long-term. For short-term investigations, time dependent material response (material creep) is considered insignificant. For long-term investigations, the time dependent response is assumed to be significant. Therefore, for long-term investigations, creep response of the refractory material should be incorporated into the material definition. Typically, if only short-term inelastic material behavior is of interest, then the combination of elastic modulus and plastic modulus mechanical material property data are required. Elastic and plastic moduli are obtained from compressive stress-strain data. An example of compressive stress-strain data for a typical 70% alumina brick is shown in Figure 3.1. Since each stress-strain curve is obtained from a test that takes only a few minutes to conduct, this data is often referred to as *static compressive stress-strain data*. In other words, the static compressive stress-strain data is short-term or time-independent material property data.

III. ASSUMPTIONS FOR ELASTIC MATERIAL BEHAVIOR

The following assumptions are made with regard to material behavior of refractories where material property descriptions are required in the analysis of refractory lining systems. The following information is not intended to be a theoretical treatise on material behavior. Rather, the following information is provided to assist in

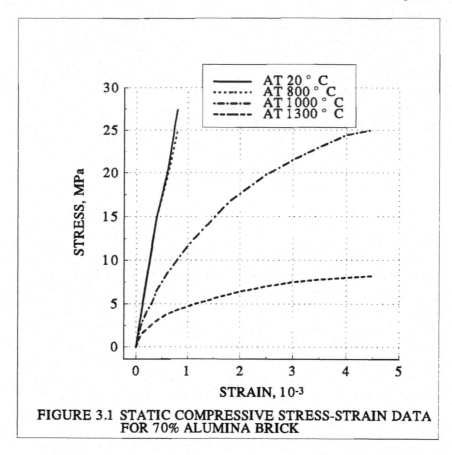

**FIGURE 3.1 STATIC COMPRESSIVE STRESS-STRAIN DATA
FOR 70% ALUMINA BRICK**

defining the most significant material properties that have been used
in refractory lining investigations described in this text which quite
accurately duplicate measured behavior of refractory lining systems.

    In general terms, it can be said that refractory materials
behave both thermally and mechanically much like other materials
used in structural systems. Typically steel, concrete and wood will
all allow heat flow through the material to varying degrees.
Through the use of heat flow calculation methodologies, the
resulting temperatures can be predicted. Also, a structure made of
these materials will deform when subjected to gravity or thermal
loads. When the loads are removed, the structure will return to the

original unloaded state, assuming excessive loading was not applied. A structure made of refractories will also exhibit the same thermal and mechanical behavior. Just as in heat flow problems, the stress-strain behavior of the refractory structure can be predicted using currently available structural analysis computer programs. In order to evaluate structural behavior of a refractory structure, however, typically the temperature-dependent stress-strain behavior of the refractory material must be defined.

For the structural analyses of refractory lining systems conducted in this text, the refractory material was assumed to behave in the classical sense. That is, the refractory material was assumed to be continuous, homogeneous and isotropic. Continuous and homogeneous meaning that the material is identical throughout the full volume of mass considered and that the smallest element of mass is identical to the largest volume of mass. Isotropic meaning that the material properties are the same in all directions. Anisotropic materials exhibit behavior that differs in various directions within the material. Orthotropic materials exhibit behavior that differs specifically in three orthogonal directions within the material.

Within the domain of elastic behavior, it is also assumed that the refractory material behaves in a linear elastic nature. For example, doubling the load results in a doubling of the stress and strain within the elastic material. References 1 and 2 are provided for those who wish to pursue a more complete understanding of classical elastic theories.

The modulus of elasticity is typically the initial tangent to the stress-strain curve as illustrated in Figure 3.2. The secant modulus is another approach to approximating the stress-strain curve for a strain range to the point of intersection between the secant tangent and the stress-strain curve. For lesser strains, however, the secant modulus underpredicts the stiffness of the material. The best approach in modeling a refractory material for structural analysis is to use the actual stress-strain curve by employing a mathematical equation that closely duplicates this curve. Most structural analysis

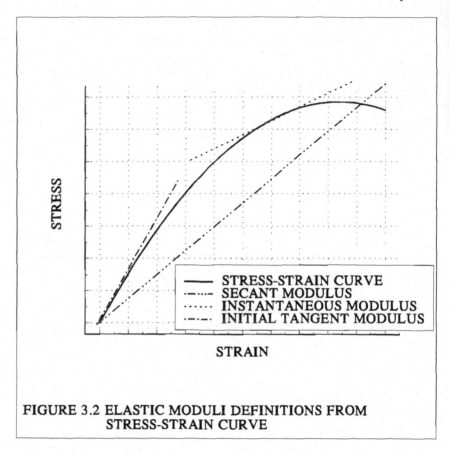

FIGURE 3.2 ELASTIC MODULI DEFINITIONS FROM
STRESS-STRAIN CURVE

programs model the material by using a series of tangent lines or
instantaneous moduli as shown in Figure 3.2. The stress-strain
curve is approximated by a series of linear parts.

IV. ASSUMPTIONS FOR INELASTIC
MATERIAL BEHAVIOR

As shown in Figure 3.1, refractory materials behave in a
linear manner at the lower temperature ranges. However, at higher
temperatures, inelastic behavior takes place as exhibited by the non-

linear stress-strain data curves for temperatures above 800°C (1500°F). Since each data curve was developed over a time period of just a few seconds or minutes, it can be concluded that the inelastic flow was primarily due to time-independent inelastic flow of the material. That is, because of the short time period, time-dependent flow or creep is not significant. The instantaneous time-dependent flow is typically called *plastic flow*. Later, it will be shown that at high temperatures significant creep flow will occur even during the short-term static compressive stress-strain tests. In the structural analysis conducted in this text, the inelastic flow of the refractory material was assumed to behave in the classical inelastic manner [3,4].

The flow curve for typical ductile material behavior is described in Figure 3.3. The form of the equation often used to define the ductile stress-strain behavior is the power equation of the form:

$$\sigma = K\varepsilon^n \qquad (3.1)$$

where K is the stress at $\varepsilon = 1.0$ and n, the strain-hardening coefficient, is the slope of a log-log plot of Equation 3.1. However, this sample equation results in considerable mathematical complexity even when the material behaves identically in all directions (isotropically).

Even for uncomplicated isotropic materials, describing the inelastic flow for three-dimensional stress states is done by using empirical relationships. Presently, there is no theoretical way of relating three-dimensional stress-state yielding to uniaxial yielding. These current empirical yield criteria are developed to match experimental observations with certain limitations. In general, the rules governing flow are as follows: the hydrostatic stress state does not cause yielding. In classical elastic theory, the hydrostatic stress is referred to as the first invariant (I_1) and defined as:

$$I_1 = \sigma_{11} + \sigma_{22} + \sigma_{33} \qquad (3.2)$$

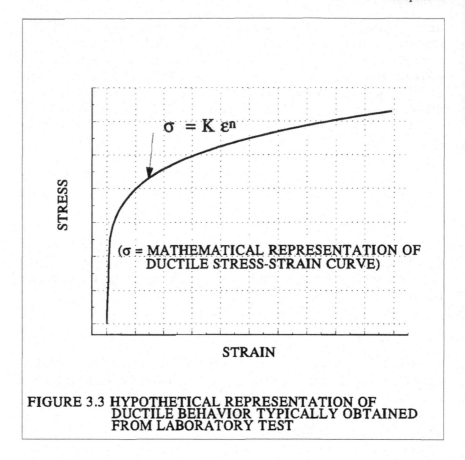

FIGURE 3.3 HYPOTHETICAL REPRESENTATION OF
          DUCTILE BEHAVIOR TYPICALLY OBTAINED
          FROM LABORATORY TEST

where the sigma values are the three principal stresses. The Von Mises, or distortion-energy, criterion for flow proposes yielding when the second invariant (J_2) exceeds a critical value (K) defined as:

$$J_2 = K^2 \qquad\qquad (3.3)$$

where $J_2 = \{(\sigma_{11} - \sigma_{22})^2 + (\sigma_{22} - \sigma_{33})^2 + (\sigma_{33} - \sigma_{11})^2\}/6$

Another popular yield criterion is the maximum shear stress, or Tresca criterion. This criterion predicts yielding when the maximum shear stress (τ_{max}) reaches the shear stress that occurs in a uniaxial tension test, defined as:

$$\tau_{max} = [\sigma_{11} - \sigma_{33}] / 2 \qquad (3.4)$$

where σ_{11} and σ_{33} are the algebraically largest and smallest principal stresses, respectively. For a uniaxial tensile test, $\sigma_{11} = \sigma_0$, and $\sigma_{33} = 0$, where σ_0 is equal to the uniaxial tensile stress, and :

$$\tau_{max} = \sigma_0 /2 \qquad (3.5)$$

Adding orthotropic relationships along with stress-dependent strain limits (small tensile stress-strain range compared to high compressive stress-strain range) adds considerably more complexity to the flow rules. Needless to say, inelastic flow problems need to be solved by computer methods. Because of the more complex refractory flow, there are very few computer programs that have the appropriate fully developed mathematical material models for refractory materials. It suffices to say computer programs with the classical approach to material behavior provide results that very closely simulate refractory material behavior.

V. USE OF CONCRETE MATERIAL CONSTITUTIVE MODELS FOR REFRACTORY MATERIALS

Refractory materials are classified as a brittle material. Basically, a material is brittle or in a brittle state if under a condition of increasing deformation it suddenly loses the ability to resist load [5]. A material is ductile or in a ductile state if the material can sustain a permanent deformation without losing its ability to resist load [5]. Later, laboratory developed compressive stress-strain data on various refractory materials which exhibit ductile flow will be discussed in detail. It will be shown that at low temperature (near ambient), refractories behave in a more brittle manner. However, at higher temperatures refractory material behaves in a more ductile manner.

Considerable amount of work has been conducted on concrete [6-57] over the past several decades, attempting to define the material behavior of concrete. The references cited here are not intended to be an exhaustive search on the subject, but they are indicative of the unsettled definition of a brittle material such as concrete with regard to the stress-strain behavior. The work on concrete has centered around several areas, including the behavior of concrete within the range of small strains. At higher strains, microcracks become more prominent and have a significant influence on the stress-strain curve. That is, according to material investigators, the apparent non-linear stress-strain behavior that takes place, as shown in Figure 3.1, is attributed primarily to microcracking. According to investigators of concrete, higher strains result in increased microcrack propagation, causing an increase in the permanent deformations. The degree or amount of microcracking has been referred to as *damage* [23,47].

Concrete has also been referred to as being orthotropic and non-linear in elastic behavior. In addition, the ratio of the various stress components for a three-dimensional stress state influences the material relationships of concrete. Needless to say, concrete material behavior is quite complex and will most likely command considerable future research attention.

Recent investigators [58] on refractory lining systems have used the material mathematical models (constitutive relationship) of concrete to evaluate the thermomechanical behavior of refractory lining systems. The use of these material models for evaluating the stress-strain behavior of refractory structures appears to be an appropriate application. However, four primary concerns are expressed here over the use of concrete constitutive material models to represent refractories.

The first concern, as cited by one concrete materials investigator [35], is that few of the material models have been systematically tested to see if they are reliable in predicting the behavior of the structure.

The second concern relates to the objective of the material model. The concrete structure is evaluated to determine when and how the concrete will fail. Refractory structures are typically restrained by steel structures and are evaluated to determine the interaction between the refractory lining and support structure. The objective of evaluating the refractory lining is not always to evaluate the failure load on the refractory lining.

The third concern is in regard to the temperature environments on concrete structures versus the temperature environments on refractory structures. Structural concrete is typically exposed to ambient temperatures. Temperatures over a few hundred degrees centigrade are most likely never experienced by concrete structures. However, refractory structures are exposed to a much higher temperature environment. For example, the steel industry, which is one of the largest consumers of refractories, exposes refractories to molten steel at nearly 1650°C (3000°F). At these higher temperatures, does the refractory truly undergo plastic flow rather than microcracking? Also will microcracks heal at the higher temperatures? At these higher temperatures there are concerns over the applicability of concrete material models to refractory materials.

The fourth and final concern is over the load environments on concrete structures versus load environments imposed on refractory structures. As expressed in the third concern, the temperature environments on refractory structures are typically much more severe than those experienced by concrete structures. The temperature environment is directly related to the load environments imposed on these two types of structures. Concrete structures are subjected to *stress-controlled* loads. That is, concrete structures primary purpose is to support gravity loads due to the weight of the structure itself plus external loads imposed on the structure. These loads create stresses within the concrete structure. If the load stresses exceed the ultimate failure stress (either ultimate tensile or compressive stress), then unlimited deformations (or strains) and a collapse of the structure would result. Refractory structures on the other hand are exposed to *strain-controlled* loads. That is, the temperature environment is the primary loading, resulting from the thermal growth restraint imposed by the steel support structure. The load is strain controlled since the thermal strains are at a set value determined by the amount of increase in temperature and the coefficient of thermal expansion. Unlike stress-controlled loading in which the stress is at a set value, strain-controlled loading results in a set value for strain. In stress-controlled loading, stress levels beyond the failure stress result in unlimited strains. In strain-controlled loading, the structure will fail if the strain exceeds the ultimate failure strain. However, the strain is limited. In other areas of material behavior, such as in the fatigue strength of materials, strain-controlled fatigue data can differ significantly from stress-controlled fatigue data. Therefore, the development of concrete material models for stress-controlled loads may not be totally applicable to refractory materials that are exposed primarily to strain-controlled loading.

VI. INFERENTIAL MATERIAL PROPERTIES

The previous discussions have provided the basic background information for defining both the thermal and mechanical material properties of refractories. ASTM test procedures [59] have been defined to assist in quantifying the mechanical behavior of refractories. The following discussions are not intended to underrate the importance of these ASTM tests. These tests provide necessary information in ranking the strength and usefulness of refractory materials. However, with regard to the structural analysis of refractory lining systems, the ASTM material data cannot be used as definitions of mechanical material property. The ASTM data is, therefore, referred to as *inferential* [60] material property information. That is, the ASTM data can be used for comparative ranking of a group of ASTM tested materials. This data, however, is not appropriate for use in the structural analysis of refractory lining systems.

A. Sonic Modulus of Elasticity (MOE)

The ASTM sonic test (C885-87) is used to evaluate the elastic modulus (young modulus or modulus of elasticity) of any general type of refractory. In the sonic test, a sample of refractory is subjected to a sonic frequency that results in the generation of a resonant frequency within the sample. Through the use of defined calculation procedures, the elastic modulus is determined. The basis for the determination of the elastic modulus by this method is that all materials with a defined geometry and defined elastic modulus will vibrate or resonate at a defined frequency. Therefore, the elastic modulus can be determined if a defined geometry of material resonates at an imposed frequency. Typically, the resonating stresses in the sample are quite small. As a result, the elastic modulus is high since the tangent to the stress-strain curve is typically steeper at the low stress level. Figure 3.4 illustrates the concept of the sonic elastic modulus versus static compressive stress-strain data. The sonic elastic modulus would be satisfactory if the actual refractory structure is exposed to a very low stress-stain environment. However, as will be shown later, refractory structures are exposed to stresses in the range of 10 to 50 MPa (several thousand psi) and that the sonic elastic modulus is not appropriate for this range of stress. Another reason for the sonic method not being useful for structural analysis is that the sonic method does not provide any information with regard to the amount of plastic flow that occurs at the higher temperatures. For these reasons, the elastic modulus data as determined by the ASTM sonic method is classified as inferential data.

The following illustrates the differences in refractories between the sonic elastic modulus and the elastic modulus calculated from static compressive stress-strain data. Figure 3.5 [61,62] compares the sonic and static elastic moduli of two similar 70% alumina brick. The sonic elastic modulus is plotted from the reported test data. The static elastic modulus was calculated using

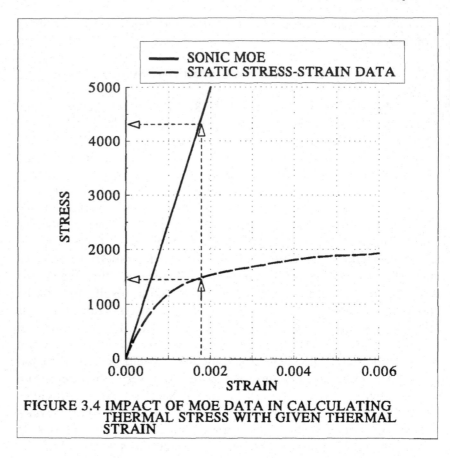

FIGURE 3.4 IMPACT OF MOE DATA IN CALCULATING
THERMAL STRESS WITH GIVEN THERMAL
STRAIN

the slope of the stress-strain curve for a strain level of one millistrain. For the high temperature stress-strain data, the higher strains result in even lower secant or instantaneous modulus values.

Figure 3.6 [63] describes another comparison of sonic and static elastic modulus data for a 58% alumina gunned castable. The static modulus was also calculated using one millistrain. As shown, the difference between the sonic and static moduli becomes greater at the higher temperatures.

Figure 3.7 [62,64] provides a comparison of sonic and static MOE data for two similar 90% alumina bricks. These data show the

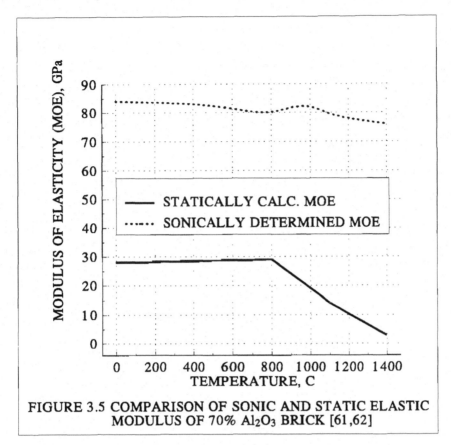

FIGURE 3.5 COMPARISON OF SONIC AND STATIC ELASTIC
MODULUS OF 70% Al_2O_3 BRICK [61,62]

sonic and static elastic modulus values are similar at low temperatures, but tend to diverge at the higher temperatures. Both the sonic and static data show a sharp reduction in the elastic modulus values at temperatures in excess of 1100 to 1200°C. The static elastic modulus was calculated using one millistrain.

Figure 3.8 [65] makes a comparison of sonic and static of an 85% Alumina castable. Just as in the previously sighted data, the static MOE has a much lower value than the sonic MOE, especially at the higher range of temperatures.

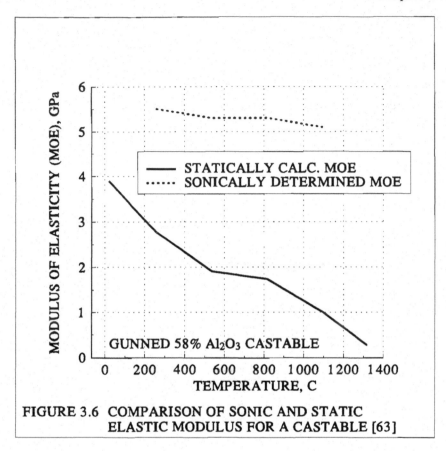

FIGURE 3.6 COMPARISON OF SONIC AND STATIC
ELASTIC MODULUS FOR A CASTABLE [63]

A final comparison of sonic and static elastic modulus of
three similar 85% alumina bricks is shown in Figure 3.9 [66]. This
work represents some of the earliest stress-strain tests on refractory
materials as well as the first comparisons of sonic and static
moduli. Just as shown in Figure 3.6, a noticeable reduction in the
sonic MOE occurs at the 1000 to 1200°C temperature range. This
behavior is also observed in some of the static MOE.

Comparison of the sonic and static MOE has also been made
in structural concrete. However, most of this work has been
conducted at ambient temperatures as structural concrete and is not
typically exposed to thermal loading. Figure 3.10 [12] describes

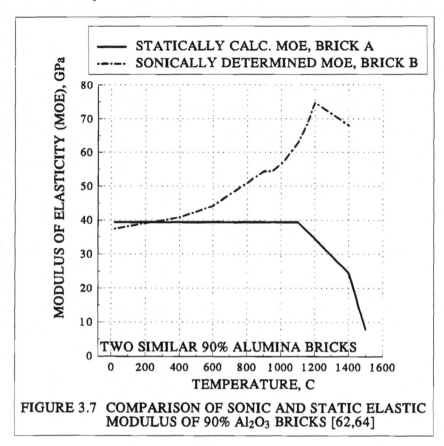

FIGURE 3.7 COMPARISON OF SONIC AND STATIC ELASTIC
MODULUS OF 90% Al_2O_3 BRICKS [62,64]

some investigative work relating the ratio of static and sonic MOE for concrete of different strength at ambient temperature. The low-strength concrete exhibits a greater ratio between static and sonic MOE than the high-strength concrete. That is, for high-strength concrete, the static and sonic MOE are nearly identical.

Based on the refractory MOE data described in Figures 3.5 through 3.9, refractory materials tend to exhibit a much greater ratio between static and sonic MOE (see Figure 3.11), especially at high temperatures. Because of the variety of refractory strengths, there is a wide range between the upper and lower limits of the static and sonic MOE ratio.

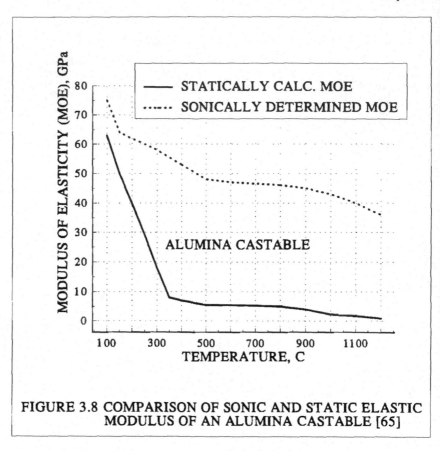

FIGURE 3.8 COMPARISON OF SONIC AND STATIC ELASTIC
MODULUS OF AN ALUMINA CASTABLE [65]

The preceding discussion regarding the differences between
sonic and static MOE is not intended to underrate the usefulness of
sonic MOE data.   Investigators [67-81] have shown that sonic
measurements can be used to rank strength and identify flaws and
other parameters relating to quality control of refractories. The sole
purpose of this discussion on sonic MOE data is to show this data
cannot be directly used in the structural analysis of refractory lining
systems.

The impact on the use of sonic versus static MOE is
demonstrated by a recent investigation [82] of steelmaking ladle
lining expansion forces. More details of this ladle investigation will

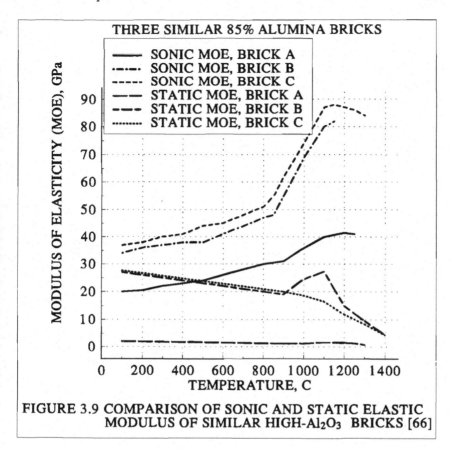

FIGURE 3.9 COMPARISON OF SONIC AND STATIC ELASTIC
MODULUS OF SIMILAR HIGH-Al$_2$O$_3$ BRICKS [66]

be provided in a later chapter. For now, the effects of sonic and
static MOE will be examined for this example. Figure 3.12
summarizes the results of the investigation. The ladle shell plate
circumferential stress was measured by several strain gages. The
strain gage test result measurement plotted represents an average of
the strain gage measurements. The strain gages measure the
resulting expansion stresses developed by the 70% alumina lining
system inside the ladle, similar to the 70% alumina brick as
characterized by the MOE data in Figure 3.5. The measurements
were made on a cold ladle during heat-up (about 7 hours with a
1000 to 1100°C hot gas temperature), during the subsequent idle
time (20-minute move of empty ladle, exposed to ambient air) and

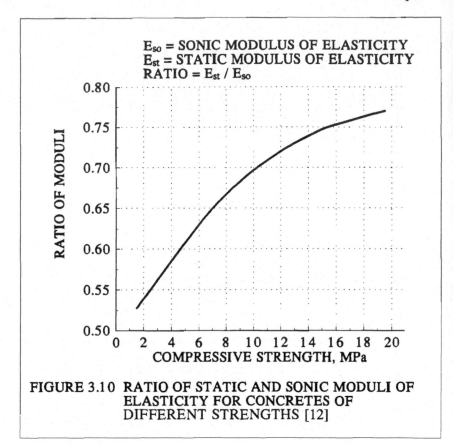

E_{so} = SONIC MODULUS OF ELASTICITY
E_{st} = STATIC MODULUS OF ELASTICITY
RATIO = E_{st} / E_{so}

FIGURE 3.10 RATIO OF STATIC AND SONIC MODULI OF
ELASTICITY FOR CONCRETES OF
DIFFERENT STRENGTHS [12]

during molten metal hold (about 1 hour at 1650°C). As shown, the strain gage results measured a maximum circumferential in the ladle sidewall shell plate stress of about 110 MPa (16,000 psi). The ladle cross section is basically cylindrical and the sidewall shell plate stress represents the summed integrated thermal expansion behavior of the refractory wall lining as reacted and restrained by the sidewall shell plate.

Two different analytical computer models were used to evaluate the thermomechanical behavior of the ladle sidewall to compare with the test results. Each model assumed a different condition for the refractory lining material properties. The

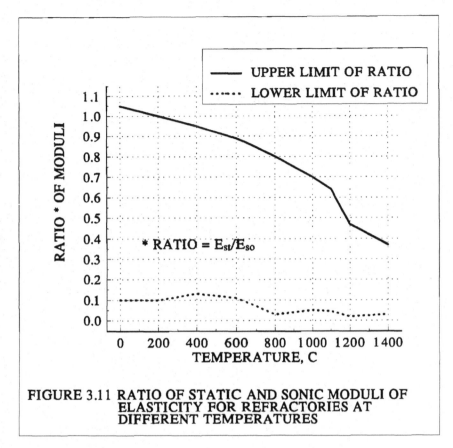

FIGURE 3.11 RATIO OF STATIC AND SONIC MODULI OF ELASTICITY FOR REFRACTORIES AT DIFFERENT TEMPERATURES

assumption of using sonic MOE data (short dash line) resulted in a predicted shell plate stress of about 280 MPa and 340 MPa during the heatup and the subsequent molten metal hold. Clearly, the sonic MOE does not result in duplicating the actual integrated mechanical response of the refractory lining.

The use of static compressive stress-strain data (Figure 3.12, dash-dot-dash line) to duplicate the elastic-plastic response of the lining material results in a greatly improved prediction of the shell plate stress. However, it should be noted that the ladle lining is a brick lining with mortar joints. The model for the material behavior assumed no mortar joints. The lining was assumed to be

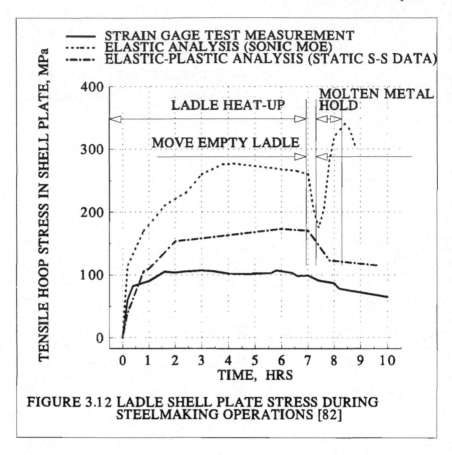

FIGURE 3.12 LADLE SHELL PLATE STRESS DURING
STEELMAKING OPERATIONS [82]

continuous. It will be shown later that the presence of mortar joints
will result in a reduction of about 30% of the stress component of
the stress-strain data and the resulting static MOE data. Therefore,
it can be concluded for the ladle study that the static compressive
stress-strain data does represent a truer representation of the
mechanical behavior of refractory material.

Another separate study [83] that identifies lining and shell
stress when sonic and static MOE are used is illustrated in Figure
3.13. The sonic MOE data on both castable and brick results in

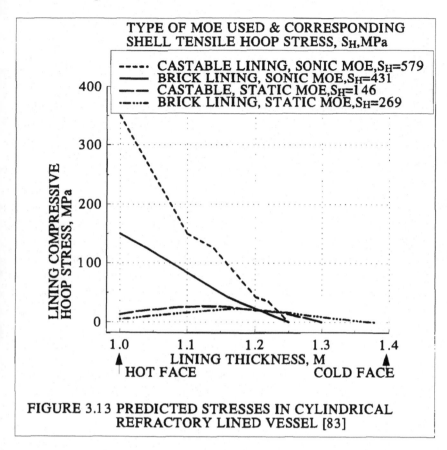

FIGURE 3.13 PREDICTED STRESSES IN CYLINDRICAL
REFRACTORY LINED VESSEL [83]

considerably greater lining stress and shell stress predictions than
when static MOE data are used. This study did not verify the
analytical results with strain gage testing, but it is expected that the
static MOE data would provide the truer stress prediction.

B. Thermal Expansion Under Load (TEUL) Test

The ASTM test (C832-89) for evaluating the thermal
expansion under load of a refractory material is classified as
inferential material property data. In this test method, the refractory

material thermal expansion is measured while simultaneously being subjected to a compressive load of at least 0.7 MPa (100 psi). The test is conducted using a one-dimensional load condition in which the amount of expansion is measured parallel to the direction of the compressive. As the temperature of the sample is continually increased, a temperature will be reached at which the onset of compressive inelastic flow will begin due to the compressive load. As the temperature is further increased the amount of plastic flow will exceed the thermal growth and noticeable slumping of the specimen will be observed. This test is then used to measure the temperature at which the onset of plastic flow occurs. Or as phrased in the ASTM C832-89 specification, Part 5, titled Significance and Use, under 5.1, "The thermal expansion under load and the 20 to 50 hour creep properties of a refractory are useful in characterizing the load bearing capacity of a refractory that is uniformly heated." These data are directly applicable for choosing refractories in blast furnace stoves and glass furnace checkers applications.

The TEUL test is used to evaluate the refractoriness under load of refractory materials, but is not useful for structural analysis for several reasons. First, this ASTM test method mixes two basic mechanical properties. Those properties are the coefficient of thermal expansion and the inelastic flow of the material. The coefficient of free thermal expansion cannot be evaluated since it is not certain how much elastic or plastic straining has taken place in the sample prior to the onset of significant plastic flow.

The second reason the ASTM TEUL data cannot be used in structural analysis is with regard to predicting the onset of inelastic flow. For most materials, the onset of plastic flow is a function of the magnitude of the stress. Typically, at low temperatures a greater stress is required to initiate plastic flow than at higher temperatures. Therefore, the onset of plastic flow may have occurred at a much lower temperature and at a much higher stress than predicted by the TEUL test. The TEUL test stress of 0.17 MPa is considerably less than the actual stress environments experienced in most refractory lining systems. For these reasons, the ASTM C832-76 TEUL test method is classified as inferential data.

Examples of TEUL test results are compared to the static compressive stress-strain curves. Figure 3.14 [84] describes the results of TEUL tests on a 98% alumina castable. A departure of the linear change in displacement occurs in a nearly proportional manner, up to a temperature of about 1250°C. Also, both the 0.2- and 0.4-MPa curves are nearly identical up to 1250°C. Above 1250°C, the percent of deformation is nearly linear with respect to the magnitude of the load. That is, the 0.4-MPa load results in a deformation departure from the no load curve of roughly twice the departure of the 0.2-MPa deformation. However, examination of the

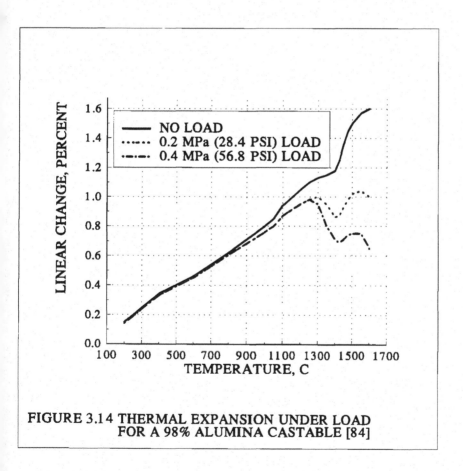

FIGURE 3.14 THERMAL EXPANSION UNDER LOAD
FOR A 98% ALUMINA CASTABLE [84]

the static compressive stress-strain data of the same castable (see Fig 3.15 [84]) shows a different trend. At a stress level of about 12 MPa, the amount of deformation (or straining) at 1370°C is about 20 times greater (0.005/0.00025) than the deformation at 1100°C. The TEUL test results do not show this proportionality of inelastic displacement between the 1370 and 1000°C deformations (as measured from the no load curve).

Additional background on TEUL testing (A 60% MgO brick) is described in Figure 3.16 [84]. This TEUL test was run using a constant 0.2-MPa load. The onset of inelastic deformation is in the range of 1550 to 1600°C. However, based on the static compressive stress-strain data (see Figure 3.17 [85]) significant inelastic

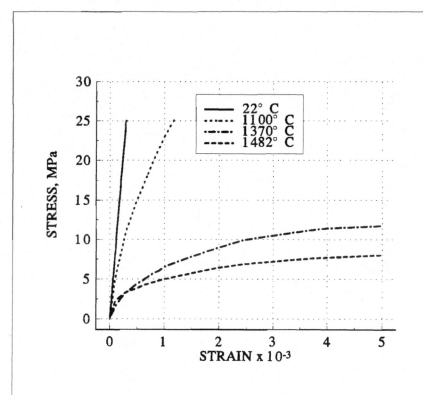

FIGURE 3.15 STATIC COMPRESSIVE STRESS-STRAIN DATA
OF 98% ALUMINA CASTABLE [85]

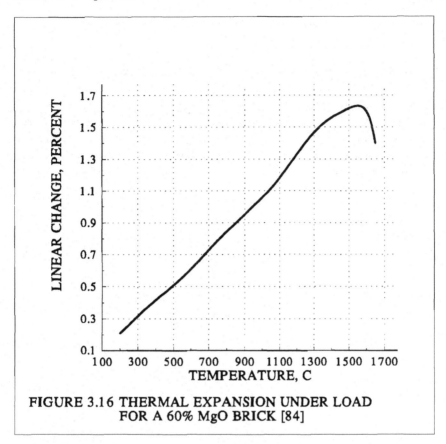

FIGURE 3.16 THERMAL EXPANSION UNDER LOAD
FOR A 60% MgO BRICK [84]

deformation occurs at temperatures in excess of approximately 1300°C.

It can be concluded, based on static compressive stress-strain data, that the TEUL test measures the onset of inelastic deformation for the very low TEUL test stress of 0.2 to 0.4 MPa. However, refractory lining systems can experience stresses of 20 to 40 MPa.

The low-stress TEUL test results do not indicate the threshold temperature at which inelastic deformation would occur at these higher stress levels.

C. Creep Test at Low Stress

The effects of creep can be defined by ASTM Test C16-81. These creep tests are typically conducted at low stress levels of about 0.20 to 2 MPa (25 to 300 psi). These stress levels, typically used to evaluate the time-dependent creep response of the refractory material, are quite low compared to actual working stress environments of refractory lining systems. Therefore, this data is questionable with regard to the applicability of this data for the analysis of refractory lining systems. For this reason, the low stress level creep data is classified as inferential data.

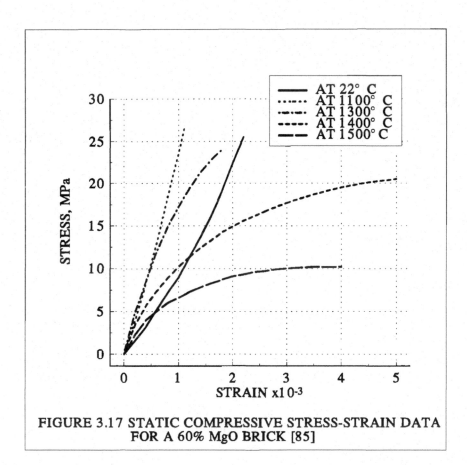

FIGURE 3.17 STATIC COMPRESSIVE STRESS-STRAIN DATA FOR A 60% MgO BRICK [85]

REFERENCES

1. Timoshenko, S. and Goodier, J. N., Theory of Elasticity, McGraw-Hill Book Co., New York, 1951.

2. Wang, C., Applied Elasticity, McGraw-Hill Book Co., New York, 1953.

3. Mendelson, A., Plasticity: Theory and Application, The MacMillan Co., New York, 1968.

4. Dieter, G. E., Materials Science and Engineering Series: Mechanical Metallurgy, McGraw-Hill Book Co., New York, 1976.

5. Jaeger, J. C. and Cook, N. G. W., Fundamentals of Rock Mechanics, John Wiley & Sons, Inc.; Chapman and Hall, London, 1976.

6. Karsan, I. D. and Jirsa, J. O., Behavior of Concrete Under Compressive Loadings, Journal of Engineering Structural Division, ASCE, Vol. 95, ST12, Proc. Paper 6935 (1969).

7. Kupfer, H., Hilsdorf, H. K. and Rusch, H., Behavior of Concrete Under Biaxial Stresses, ACI Journal, Vol. 66, No. 8 (1969).

8. Baldwin, R. and North, M. A., Stress-Strain Curves of Concrete at High Temperatures--A Review, Fire Research Note No. 785, Ministry of Technology and Fire Offices Committee, Oct., 1969.

9. Buyukozturk, O., Stress-Strain Response and Fracture of A Model of Concrete in Biaxial Loading, Ph.D. Thesis, Cornell University, 1970.

10. <u>Reinforced Concrete Engineering</u>, Bresler, B., Editor, John Wiley & Sons, New York, 1974.

11. Harmathy, T. Z., Thermal Properties of Concrete at Elevated Temperatures, JMLSA, Vol. 5, No. 1, March, pp. 47-74 1970.

12. Neville, A. M., <u>Properties of Concrete</u>, John Wiley & Sons, New York, 1973.

13. Green, S. J. and Swanson, S. R., Static Constitutive Relations for Concrete, Technical Report No. AFWL-TR-72-2, Terra-Tec, Inc., Salt Lake City, Utah (1973).

14. William, K. J. and Warnke, E. P., Constitutive Model for the Triaxial Behavior of Concrete, Proceedings IABSE, Borgamo, Italy, (1974).

15. Romstad, K. M., Taylor, M. A., and Herrmann, L. R., Numerical Biaxial Characterization for Concrete, Journal of Engineering Mechanics Division., ASCE, Vol. 100, EM5, Proc. Paper 10879 (1974).

16. Băzant, Z. P., Instability, Ductility, and Size Effect in Strain-Softening Concrete, Journal of Engineering Mechanics Division., ASCE, Vol. 102, EM2, Proc. Paper 12042 (1976).

17. Darwin D. and Pecknold, D. A., Non-Linear Biaxial Stress-Strain Law for Concrete, Journal of Engineering Mechanics Division., ASCE, Vol. 103, EM2, Proc. Paper 12839 (1977).

18. Cedolin, L., Crutzen, Y. R. J., and Dei Pole, S., Triaxial Stress-Strain Relationship for Concrete, Journal of Engineering Mechanics Division, ASCE, Vol. 103, EM3, Proc. Paper 12969 (1977).

19. Johnson, W. and Mellor, P. B., Engineering Plasticity, Van Nostrand Reinhold Company, London, England, 1978.

20. Kotsovos, M. D. and Newman, J. B., A Mathematical Description of the Deformational Behavior of Concrete Under Complex Loading. Magazine of Concrete Research, Vol. 31, No. 107 (1979).

21. Kotsovos, M. D. and Newman, J. B., A Mathematical Description of Deformational Behavior of Concrete Under Generalized Stress Beyond Ultimate Strength, ACI Journal, Vol. 77, No. 5 (1980).

22. Braiden, P. M., The Development of Rational Design Criteria for Brittle Materials, Materials in Engineering, Vol. 2, Dec. (1980).

23. Krajcinovic, D. and Fonseka, G. U., The Continuous Damage Theory of Brittle Materials: Part I--General Theory; Part 2--Uniaxial and Plane Response Models, Journal of Applied Mechanics, Vol. 48 (1981).

24. Băzant, Z. P., et al., Normal and Refractory Concrete for LMFBR Applications; Volume 2: Evaluation of Concretes for LMFBR Applications, EPRI Report NP-2437 (1982).

25. Scavuzzo, R., et al., Simple Formulation of Concrete Response to Multiaxial Load Cycles, Proceedings of The International Conference on Constitutive Laws for Engineering Materials: Theory and Application (1983).

26. Fardis, M. N., Alibe, B., and Tassoulas, J. L., Monotonic and Cyclic Constitutive Law for Concrete, Journal of the

Engineering Mechanics Division, ASCE, Vol. 109, No. EM2, Proc. Paper 17871 (April 1983).

27.   Kotsovos, M. D., Concrete--A Brittle Fracture Material, Material and Structures, RILEM (in press).

28.   Ahmad, S. H., et al., Orthotropic Model of Concrete for Triaxial Stresses, Journal of Structural Engineering, ASCE, Vol. 112, No. 1, Paper 20290 (1986).

29.   Cedulin, L., Sandro, D. P., and Ivo, I., Tensile Behavior of Concrete, Journal of Engineering Mechanics, ASCE, Vol. 113, No. 3, Paper 21363 (1987).

30.   Zubelewicz, A. and Băzant, Z. P., Constitutive Model with Rotating Active Plane and True Stress, Journal of Engineering Mechanics, ASCE, Vol. 113, No. 3, Proc. Paper 21346 (1987).

31.   Rots, J. G. and deBorst, R., Analysis of Mixed-Mode Fracture in Concrete, Journal of Engineering Mechanics, ASCE, Vol. 113, No. 11, Paper 21979 (1987).

32.   Graves, R. H. and Derucher, K. N., Interface Smeared Crack Model Analysis of Concrete Dams in Earthquakes, Journal of Engineering Mechanics, ASCE, Vol. 113, No. 11, Paper 21954 (1987).

33.   Ohtani, Y. C. and Chen, W.-F., Hypoelastic--Perfectly Plastic Model for Concrete Materials, Journal of Engineering Mechanics, ASCE, Vol. 113, No. 12, Paper 22012 (1987).

34.   Yong, Y.-K., et al., Behavior of Laterally Confined High-Strength Concrete Under Axial Loads, Journal of Structural Engineering, ASCE, Vol. 114, No. 2, Paper 22199 (1988).

35. Balakrishnan, S. and Murray, D. W., Concrete Constitutive Model for NLFE Analysis of Structures, Journal of Structural Engineering, ASCE, Vol. 114, No. 7, Paper 22569 (1988).

36. Băzant, Z. P. and Prat, P. C., Microplane Model for Brittle-Plastic Material: I. Theory, Journal of Engineering Mechanics, ASCE, Vol. 114, No. 10, Paper 22823 (1988).

37. Băzant, Z. P. and Prat, P. C., Microplane Model for Brittle-Plastic Material: II. Verification, Journal of Engineering Mechanics, ASCE, Vol. 114, No. 10, Paper 22824 (1988).

38. Yankelevsky, D. Z. and Reinhardt, H. W., Uniaxial Behavior of Concrete in Cyclic Tension, Journal of Structural Engineering, ASCE, Vol. 115, No. 1, Paper 23126 (1989).

39. Mazars, J. and Pijaudier-Cabot, G., Continuing Damage Theory, Application to Concrete, Journal of Engineering Mechanics, ASCE, Vol. 115, No. 2, Paper 23187 (1989).

40. Krevzer, H. and Bury, K. V., Reliability Analysis of Mohr Failure Criterion, Journal of Engineering Mechanics, ASCE, Vol. 115, No. 3, Paper 23236 (1989).

41. Yoshikawa, H., Wu, Z. and Tanabe, T., Analytical Model for Shear Slip of Cracked Concrete, Journal of Structural Engineering, ASCE, Vol. 115, No. 4, Paper 23347 (1989).

42. Pramono, E. and William, K., Fracture Energy-Based Plasticity Formulation of Plain Concrete, Journal of Engineering Mechanics, ASCE, Vol. 115, No. 6, Paper 23533 (1989).

43. Franchi, A. and Genna, F., Self-Adaptive Model for Structural Softening of Brittle Materials, Journal of

Engineering Mechanics, ASCE, Vol. 115, No.7, Paper 23704 (1989).

44. deBoer, R. and Dresenkamp, H. T., Constitutive Equations for Concrete in Failure State, Journal of Engineering Mechanics, ASCE, Vol. 115, No. 8, Paper 23723 (1989).

45. Shih, T. S., Lee, G. C. and Chang, K. C., On Static Modulus of Elasticity of Normal-Weight Concrete, Journal of Structural Engineering, ASCE, Vol. 115, No. 10, Paper 23979 (1989).

46. Otter, D. E., and Naaman, E., Model for Response of Concrete to Random Compressive Loads, Journal of Structural Engineering, ASCE, Vol. 115, No. 11, Paper 24061 (1989).

47. Suaris, W., Ouyang, C. and Fernando, V. M., Damage Model for Cyclic Loading of Concrete, Journal of Engineering Mechanics, ASCE, Vol. 116, No. 5, Paper 24617 (1990).

48. Yamaguchi, E. and Chen, W.-F., Cracking Model for Finite Element Analysis of Concrete Materials, Journal of Engineering Mechanics, ASCE, Vol. 116, No. 6, Paper 24721 (1990).

49. Dahlblom, O. and Ottosen, N. S., Smeared Crack Analysis Using Generalized Fictitious Crack Model, Journal of Engineering Mechanics, ASCE, Vol. 116, No. 1, Paper 24207 (1990).

50. Liaw, B. M., et al., Improved Non-Linear Model for Concrete Fracture, Journal of Engineering Mechanics, ASCE, Vol. 116, No. 2, Paper 24368 (1990).

51. Berthelot, J. M. and Robert, J. L., Damage Evaluation of Concrete Test Specimens Related to Failure Analysis, Journal of Engineering Mechanics, ASCE, Vol. 116, No. 3, Paper 24430 (1990).

52. Du, J. et al., Direct FEM Analysis of Concrete Fracture Specimens, Journal of Engineering Mechanics, ASCE, Vol. 116, No. 3, Paper 24431 (1990).

53. Yazdani, S. and Schreyer, H. L., Combined Plasticity and Damage Mechanics Model for Plain Concrete, Journal of Engineering Mechanics, ASCE, Vol. 116, No. 7, Paper 24822 (1990).

54. Liaw, B. M., et al., Fracture-Process Zone for Mixed-Mode Loading of Concrete, Journal of Engineering Mechanics, ASCE, Vol. 116, No. 7, Paper 24830 (1990).

55. Băzant, Z. P., et al., Random Particle Model for Fracture of Aggregate or Fibet Composites, Journal of Engineering Mechanics, ASCE, Vol. 116, No. 8, Paper 24923 (1990).

56. Fanella, D. A., Fracture and Failure of Concrete in Uniaxial and Biaxial Loading, Journal of Engineering Mechanics, ASCE, Vol. 116, No. 11, Paper 25184 (1990).

57. Yamaguchi, E. and Chen, W.-F., Microcrack Propogation Study of Concrete Under Compression, Journal of Engineering Mechanics, ASCE, Vol. 117, No. 3, Paper 25598 (1991).

58. Chen, E.-S. and Buyokozturk, O., Behavior of Refractory Linings for Slagging Gasifiers, MIT Report R84-07, MIT-CE-7862-02, U.S. Dept. of Energy, June (1984).

59. 1993 Annual Book of ASTM Standards, Vol. 15.01, Refractories; Carbon and Graphite Products; Activated Carbon, ASTM, Philadelphia, PA, 1993.

60. Astbury, N. F., Deformation and Fracture, Trans. Brit. Ceram. Research Association, pp. 1-7, January (1969).

61. Kaiser Refractories (National Refractories & Minerals Corp.) Technical Data, Krial 70-HS (70% Al_2O_3 Brick) Data Sheet and Static Compressive Stress-Strain Data (Circa 1990).

62. Harbison-Walker Refractories Company, H-W 33-66 (70% Al_2O_3 Brick), Korundal XD (90% Al_2O_3 Brick), Sonic Elastic Modulus Data Versus Temperature, Project 5503 (1967).

63. Alder, W. R., Stress/Strain Testing of Refractories for the Hydrocarbon Processing Industry, Kaiser Refractories, Center for Technology (National Refractories & Minerals Corp.) (Circa 1983).

64. Kaiser Refractories (National Refractories & Minerals Corp.) Technical Data, Kricor Data Sheet and Static Compressive Stress-Strain Data (Circa 1981).

65. Palin, F. T. and Padgett, G. C., Thermo-Mechanical Behavior of Refractory Castable Linings, Brit. Ceram. Res. Assn., Technical Note No. 320, Refractories and Industrial Ceramics Division, March (1981).

66. Padgett, G. C., Cox, J. A., and Clements, J. F., Stress/Strain Behavior of Refractory Materials at High Temperatures, Trans. Brit. Ceram. Soc., 68 (2), pp. 63-72 (1969).

67. Davis, W. R., Measurement of Mechanical Strength of Refractory Materials by a Non-Destructive Method, Research

Paper No. 395, Brit. Ceram. Res. Assn., Stoke-On-Trent, England (1955).

68. Davis, W. R. and Clements, J. F., Sonic Testing, Proceedings, Mechanical Properties of Non-Metallic Brittle Materials, Interscience, New York, pp. 203-210 (1958).

69. Vassiliou, B. E. and Baker, C. J. W., A Simple Laboratory Method for Determinng the Modulus of Elasticity, Brit. Ceram. Res. Assn., Stoke-On-Trent, England, Research Paper No. 346 (1956).

70. Semler, C. E., Sonic Testing of Refractories, Abstract in Am. Ceram. Soc. Bull., Vol. 57, No. 3, p. 366 (1978).

71. Judd, M. S. et al., Improved Reliability of Fused Silica Pouring Tubes for Continuous Casting with Non-Destructive Inspection, Abstract in Am. Ceram. Soc. Bull., Vol. 58, No. 3, p. 317 (1979).

72. Whittemore, D. S., Sonic Velocity Measurement on Pouring Pit Refractories, Abstract in Am. Ceram. Soc. Bull, Vol. 58, No. 3, p. 317 (1979).

73. Dunworth, B., Pulse Ultrasonics in Process Control, Abstract in Am. Ceram. Soc. Bull, Vol. 58, No. 8, p. 800 (1979).

74. Miller, W., Refractory Evaluation with Pulse Ultrasonics, Abstract in Am. Ceram. Soc. Bull., Vol 58, No. 8, p. 800 (1980).

75. Lawlar, J. B., Ross, R. H. and Ruh, E., Nondestructive Ultrasonic Testing of Fireclay Refractories, Am. Ceram. Soc. Bull., Vol. 60, No. 7, pp. 713-718, (1981).

76. Semler, C. E., Non-Destructive Ultrasonic Evaluation of Refractories, Inter-ceram, No. 5, pp. 485-488 (1981).

77. Canelli, G. and Monti, F., Ultrasonic Inspection of Slide Gate Valve Components, Refr. J., Vol. 2, Mar.-Apr., pp. 11-12 (1981).

78. Petit, J., Relation entre le module d'elasticite et les proprietes physiques des materiaux refractaires. Application aux briques de poches d'acierie, Revue de Metall., Aug.-Sept., pp. 657-668 (1983).

79. Kawai, H. et al., Detectability of Natural Flaws in Sintered SiC by Ultrasonic Technique, Preprint of International Symposium on Ceramic Components for Engines, Hakone, Japan (1983).

80. Russell, R. O. and Morrow, G. D., Sonic Velocity Quality Control of Steel Plant Refractories, Am. Ceram. Soc. Bull., Vol. 63, No. 7, pp. 911-914 (1984).

81. Aly, F. and Semler, C. E. Prediction of Refractory Strength, Using Non-Destructive Sonic Measurements, Am. Ceram. Soc. Bull., Vol. 64, No. 7, pp. 1555-1558 (1985).

82. Schacht, C. A., Improved Mechanical Material Property Definition for Predicting The Thermo-Mechanical Behavior of Refractory Linings of Teeming Ladles, Journal of the Amer. Ceram. Soc., Vol. 76, No. 1, pp. 202-206 (1993).

83. Padgett, G. C. and Palin, F. T., Engineering of Refractory Structures. Brit. Ceram. Research Ltd., Stoke-on-Trent, England (Circa 1991).

84.   National Refractories & Minerals Corporation, Technical Material Data on Laboratory Made Sample, SC-147, 98% Al_2O_3 Castable (Circa 1981).

85.   Stett, M. A., Measurement of Properties for Use with Finite Element Analysis Modeling, Amer. Ceram. Soc., Proceedings Ceramic and Engineering Science, ISSN 0196-6219, Vol. 7, No. 1-2, pp. 196-208 (1986).

4
ASTM Strength Tests

I. INTRODUCTION

The uniaxial strength of refractory materials differs considerably with regard to ultimate compressive strength versus ultimate tensile strength. The ultimate compressive strength of refractories is typically two to ten times greater [1,2] than the ultimate tensile strength. However, this is not a unique characteristic of refractory materials. Graphite, structural concrete, rock and other brittle materials exhibit similar ranges between tensile and compressive strengths.

Ultimate tensile and compressive strengths are estimated by ASTM test methods. The following discussions address the ultimate strengths of refractories, and the interpretation of the ASTM test results along with other parameters that surround the development of the ultimate strength. With regard to ultimate strength, it should be noted that the ASTM tests are used to estimate the ultimate strength of refractories. It is recognized in the material science community of refractory materials [1] that the following ASTM tensile strength tests to be discussed do not evaluate the true ultimate tensile strength, but rather provide an estimate of the ultimate tensile strength of refractories.

It should also be noted that the following ASTM tensile and compressive strength tests are load controlled tests. Therefore, only the specimen geometry is required to determine the resulting stress within the specimen. Needing only the geometry to evaluate the failure stress provides no information on the non-linear mechanical material behavior. It will be shown in the following discussion that the refractory modulus of elasticity (MOE) is not linear with respect to the sign of stress (compressive or tensile) and the magnitude of stress. Therefore, the specimen modulus of elasticity, in addition to the specimen geometry, is needed to define the estimated ultimate tensile stress in some of the ASTM tests.

II. ASTM TESTS ON TENSILE STRENGTH OF REFRACTORIES

The ASTM C1099-92, C583-80 and C491-85 modulus of rupture (MOR) tests [2] are used to estimate the ultimate tensile strength (f'_t) of refractories. C491-85 is an MOR test for air-setting plastic refractories but was discontinued in 1993. These tests are used to quantify the modulus of rupture of refractory materials. There are several types of tests to estimate the ultimate tensile strength of brittle materials such as refractories. The three most popular are defined as the modulus of rupture test (as identified above), the splitting tensile strength (identified here as STS) test (ASTM C496-69 [3]) and the direct, or uniaxial, tensile pull (identified here as DTP) test (ASTM C190-58, discontinued [4]). The STS test is also referred to as the Brazil test. Each of these three tests can be used to evaluate the ultimate tensile strength of brittle materials. The more commonly used test to estimate the ultimate tensile strength in refractories is the MOR test. According to investigators of refractory material tensile strength, the MOR test is more reliable and, therefore, has become, the more popular test used throughout the industry.

Comparison of the test results for the three testing methods applied to concretes show the MOR test consistently predicts higher ultimate tensile strength values than that of the STS and DTP tests. Typical ratios of estimated ultimate tensile strength data on the same material for the three tests [Chap. 2, ref. 10] are:

$$f'_{t,MOR}/F'_{t,MOR} = 1.41 \text{ to } 1.83$$

And the $f'_{t,MOR}$ is, on the average, 50% higher than $f'_{t,DTP}$. According to other investigators [Chap. 2, ref. 12], ultimate tensile strength investigations on concrete indicates the $f'_{t,STS}$ is typically 5 to 12% higher than $f'_{t,DTP}$ for the same material. The latter reference also indicates that $f'_{t,MOR}$ values are twice that of $f'_{t,DTP}$ values for concrete materials.

The variation in tensile strength as a result of the type of tensile test chosen has not been fully understood. Regardless of the test used, Griffith [5] contends that the measured strength is always lower than the theoretical molecular cohesion strength because of the presence of small cracks and flaws in the specimen. The orientation of the flaws and the locations of the flaws (internal or surface) relative to the applied stress influence the stress concentration effects caused by the flaws. As the applied load, and resulting equilibrating stress, is increased, the number of flaws tend to increase along with the size of the flaws. The most critical flaw (flaw located in region of greatest equilibrating stress) may be submicroscopic in size prior to the specimen loading. Griffith's theory of tensile strength, however, does not totally explain the consistent difference in refractory tensile strength values between the three types of tensile strength tests. Some investigators suggest that inherent eccentricity of loading in the DTP test results in this test developing lower estimated ultimate tensile strength data. That is, there may be an unaccounted bending moment introduced in the sample that results in an additional unaccounted tensile stress. This additional bending stress adds to the tensile stress from the direct pull load resulting in an effectively lower failure load at tensile failure. This would suggest that the DTP test has more inherent eccentricity than the STS and MOR tests.

According to Duckworth [1], the MOR single-point loading (see Figure 4.1) method in ASTM C583-92 results in a maximum bending stress (f_t) in the specimen at a single point. Low probability exists for the most dangerous (or most critical) flaw to be located at the region of the peak bending stress with two point loading (see Figure 4.2), in which the peak bending stress is distributed over a significant portion of the specimen length.   Therefore, a higher probability exists for having a most dangerous flaw in the region of greatest equilibrating stress. Duckworth identifies factors relating to the geometry of the specimen and testing device that may lower MOR values. These factors are described as (1) torsion introduced in the specimen by non-uniform loading across the breadth of the specimen, (2) poor specimen corners due to fabrication of the specimen cross section (effectively resulting in a lower moment of

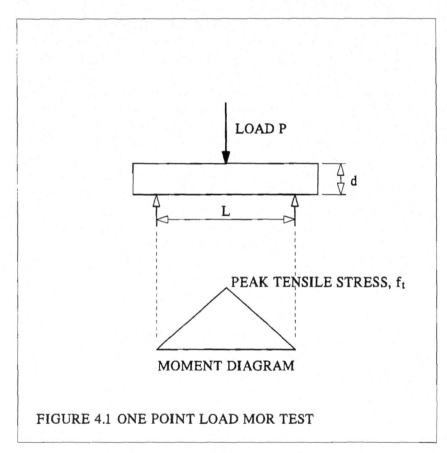

FIGURE 4.1 ONE POINT LOAD MOR TEST

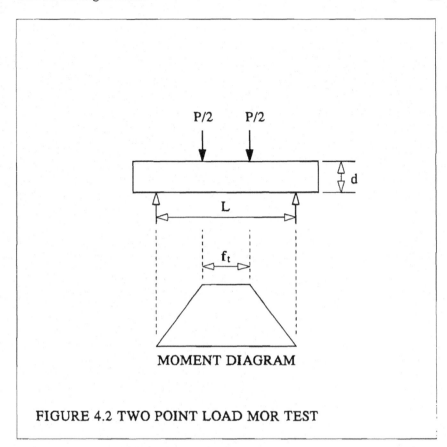

FIGURE 4.2 TWO POINT LOAD MOR TEST

inertia), (3) curvature of the specimen along the length, (4) unequal loading at the two load points and (5) localized stresses in the specimen at load points.

The STS tensile stress test results are reportedly similar to the DTP test results. However, Neville [Chap. 2, ref. 12] indicates that the STS test results on concrete are consistently 5 to 12% higher than the DTP test results. Bresler [Chap. 2, ref. 10] indicates STS concrete test results are 5% lower than the DTP concrete test results. He attributes this to the biaxial compressive stress state developed during testing of the STS sample but does not elaborate on any details.

Tensile strength tests on various types of rock have shown similar relationships as to those reported on rocks and concrete (see Tables I and II) [6,7].

Duckworth [1], Jaeger [6] and Schacht [8] have shown that the stress-dependent modulus of elasticity of the material can have a significant influence in estimating the ultimate tensile strength predicted by the ASTM MOR tests. Schacht, however, goes on to explain the significance of the stress sign-dependent MOE on the STS test and use finite element methods to describe the significance of MOE on predicting tensile strength [8]. Most of the data on the stress dependency of refractory modulus of elasticity shows the

Table I

Comparison of Tensile Strengths (psi)*

Method		Rock	
	Marble, Carrara	Sandstone, Gosford	Trachyte, Bowral
Uniaxial tension	1,000	520	1,990
Brazilian (15 arc)	1,265	540	1,740
Hollow Brazilian ($R_2/R_1 = 2$)	2,500	1,200	3,500
Bending (3 point)	1,710	1,140	3,659

*From Ref. 6

Table II

Variability of Results of Tests on Tensile
Strength of Concrete*

Type of Test	Mean Strength MN/m^2 lb/in.^{2}		Standard Dev. Within Batches MN/m^2 lb/in.^{2}		Coeff. of Variation Per Cent
Splitting test	2.79	405	0.14	20	5
Direct tensile test	1.90	275	0.13	29	7
MOR	4.17	605	0.25	36	6
Compression cube test	41.23	5,980	1.43	207	3.5

*From Chap. 3, Ref. 12

tensile MOE to be less than the compressive MOE. The available literature regarding stress-dependent compressive and tensile static stress-strain data is rather limited. Neville suggests that for structural concrete the modulus of elasticity in tension is less than the modulus of elasticity in compression [Chap. 2, ref. 12]. Jaeger [7] expresses similar characteristics for rock material. Other material investigators [7,9,10] also show the tensile modulus of elasticity is less than the compressive modulus of elasticity.

III. CURSORY ANALYTICAL INVESTIGATIONS OF ASTM TENSILE TESTS USING CLASSICAL MATHEMATICS

The following discussion addresses the actual stress states developed in the types of specimens used in three different ASTM tensile test methods. The objective of identifying the actual stress states is to better assist in explaining the differences in the estimated ultimate tensile stress due to a stress sign- (compressive or tensile stress) dependent MOE.

It will be shown that the stress states that exist in the three ASTM tensile tests due to the stress sign-dependent MOE result in three distinctively different assumptions regarding material behavior. With regard to the DTP test, the stress state is quite simple. Here, the principal tensile stress is basically a one-dimensional tensile stress. No compressive stresses exist to complicate the issue.

The MOR test is basically a one-dimensional stress state. However, since internal bending strength is the mode of resisting the externally applied load, both compressive and tensile one-dimensional stresses are developed in the beam specimen. Compressive stress exists in the top portion and tensile stress exists in the bottom portion of the beam specimen. For MOE data that is not dependent on stress, the beam neutral axis will be at the beam midheight. With these regions of compressive and tensile stress in the specimen, an MOE that is dependent on the sign of the stress will cause the beam neutral axis to shift from the midheight location.

The STS test specimen stress state is unique compared to the DTP and MOR specimen stress states. The STS is basically a two-dimensional stress state. In addition, one dimension of the stress state (parallel to the load direction) is compressive throughout the full region of the specimen, while the second dimension of the stress state (perpendicular to the load direction) is tensile throughout the full region. In this case, the elastic modulus is different in the two principal directions (orthotropic) over the full region of the specimen.

The three different ASTM tensile tests result in three distinct differences in material behavior when the MOE is a function of tensile and compressive stress states. The following discussions provide more detailed information on effects of stress sign-dependent MOE.

A. DTP Test Method

The direct tensile pull specimen is the only specimen that is exposed to a pure tensile stress state in the region of expected fracture. As a result, the stress state is uncomplicated compared to the stress states in MOR specimen and the STS specimen. Because of the uniaxial tensile stress state, the stress-dependent modulus of elasticity plays no role in influencing the estimated tensile stress state. The estimated ultimate tensile stress is evaluated by simply dividing the ultimate tensile load by the cross-sectional area of the specimen. Also, as previously noted, the DTP test is a stress-controlled test and since only a pure tensile stress is developed, only the specimen geometry is required for calculating the estimated tensile strength.

B. MOR Test Method (Linear MOE)

The MOR specimen stress state is a bit more complicated than the DTP specimen stress state. The MOR specimen is a simple supported beam. The internal equilibrating bending moment is developed along the length of the specimen, as previously illustrated in Figure 4.1. In some instances, the MOR test is conducted using two point loading as shown in Figure 4.2. The following discussion can be addressed to either the one or two point loading. However, for our purposes here, the one point loading is used.

The moment in the specimen increases linearly from zero value at the two end support points to a maximum at the midspan load position. At midspan the internal equilibrating moment (M_{INT}) for the one point loading is equal to the external moment (M_{EXT}):

$$M_{INT} = M_{EXT} = (P/2)(L/2) = PL/4 \qquad (4.1)$$

For materials in which the modulus of elasticity is not influenced by the sign (compressive or tensile) or magnitude of the stress state, the tensile bending stress (f_t) at the bottom of the beam is equal and opposite to the top compressive stress (f_c) for any vertical plane of interest. The bending stress is calculated by dividing the moment by the section modulus (S). For the purposes of this discussion, it is assumed the MOR specimen has a rectangular cross section and the section modulus (S) is:

$$S = bd^2/6 \qquad (4.2)$$

where b is the beam width and d the beam depth. Therefore:

$$f = M/S = (PL/4)(6/bd^2) = 1.5\ PL/bd^2 \qquad (4.3)$$

in which the + notation is used to define the bottom tensile stress and the - notation is used to define the top compressive stress. In this case, the neutral axis of bending in the cross section of the beam is located at the beam midheight. That is, the neutral axis is located at d/2, measured either from the top or bottom of the beam. The midheight neutral axis (NA) is shown by the dot dash line in Figure 4.3. These relationships can be made with regard to the MOR test if the stress-strain relationships are linear. That is, neither the sign nor the magnitude of the stress influence the value of the MOE.

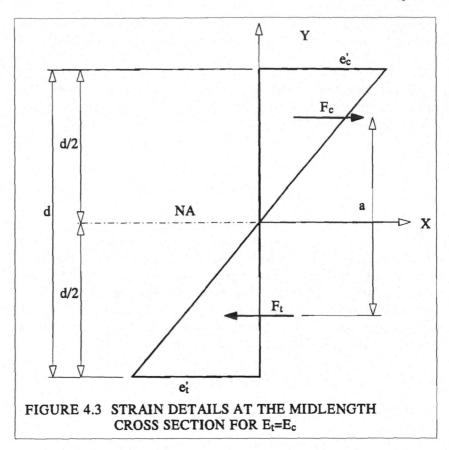

FIGURE 4.3 STRAIN DETAILS AT THE MIDLENGTH
CROSS SECTION FOR $E_t = E_c$

C. MOR Test Method (Stress Sign-Dependent MOE)

For refractories in which the MOE is dependent on the sign of the stress, the tensile MOE is in most cases less than the compressive MOE. As described in Figure 4.4, the neutral axis will not be located at the midheight, but above the beam midheight, as shown by the following equations. It is assumed that the MOE is non-linear with respect to the sign of the stress but constant with respect to the magnitude of either the compressive or tensile stress, as illustrated in Figure 4.5.

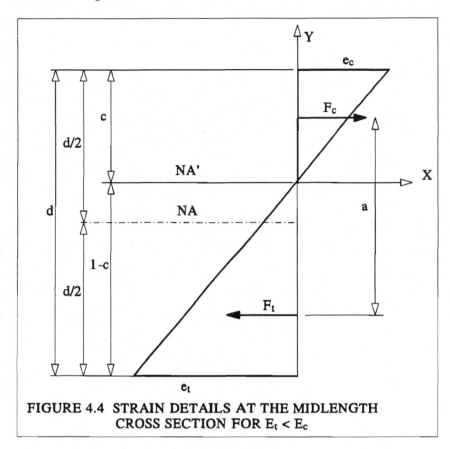

FIGURE 4.4 STRAIN DETAILS AT THE MIDLENGTH
CROSS SECTION FOR $E_t < E_c$

Since the tensile MOE is less than the compressive MOE, the beam will require a larger volume of tensile strain than compressive strain to allow internal force equilibrium to occur. The basic assumptions [11,12] regarding beam behavior are (referring to Figure 4.4) that internal equilibrium requires that F_c equal F_t. It still holds true that plane sections prior to bending remain plane sections after bending occurs. Therefore, bending strains increase linearly from the neutral axis. The upper and lower triangles in Figure 4.4 represent the upper compressive and lower tensile strain states in the beam. Multiplying these strains by the appropriate elastic moduli (see Figure 4.5) will result in the compressive and tensile stresses.

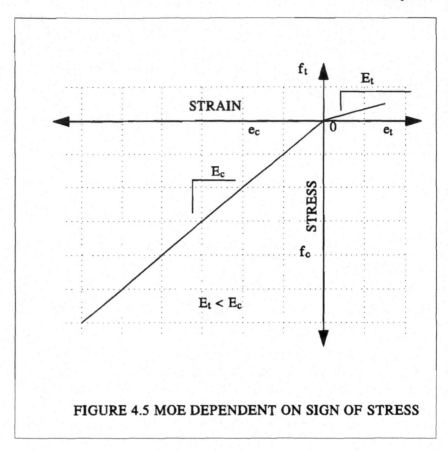

FIGURE 4.5 MOE DEPENDENT ON SIGN OF STRESS

The linear relationship between stress and strain can be defined algebraically as:

$$f = Ae \qquad (4.4)$$

For the compressive stress (f_c), the constant term A is the compressive MOE times the compressive strain (e_c), or:

$$f_c = E_c e_c \qquad (4.5)$$

Likewise, for the tensile stress (f_t):

$$f_t = E_t e_t \qquad (4.6)$$

Therefore, the internal compressive force (F_c) is represented by the volume of the stress diagram, which is equal to the compressive strain multiplied by the compressive MOE:

$$F_c = f_c cd/2 = E_c(e_c cd/2) \tag{4.7}$$

where C is the parameter that locates the neutral axis. Likewise for the internal tensile force:

$$F_t = f_t(1 - c)d/2 \; E_t[e_t(1 - c)d]/2 \tag{4.8}$$

These internal forces are located at the center of gravity of their appropriate strain diagrams. Because of the linear relationship between stress and strain, the center of gravity of the stress volumes (and location of internal forces) is identical to the center of gravity of the strain volumes. The distance separating these forces is two-thirds the beam depth. The internal moment can be defined as:

$$M_{INT} = F_t \, (2d/3) \tag{4.9}$$

From Equation 4.1:

$$F_t(2d/3) = PL/4 \tag{4.10}$$

Rearranging:

$$F_t = 3PL/8d \tag{4.11}$$

Equation 4.8 can be defined, using Equation 4.6 as:

$$F_t = f'_t(1 - c)bd/2 \tag{4.12}$$

in which f'_t is the estimated ultimate tensile stress. From Equations 4.11 and 4.12:

$$f'_t(1 - c)bd/2 = 3PL/8d$$

$$f'_t = (1 - c)(3PL/4bd^2) \tag{4.13}$$

Since plane sections remain plane:

$$e'_t c = e_c(1 - c)$$

or:

$$e'_t = e_c(1 - c)/c \qquad (4.14)$$

where e'_t is the estimated ultimate tensile strain.

From the assumption of equilibrium:

$$F_t = F_c \qquad (4.15)$$

Therefore, from Equations 4.7 and 4.8 and from internal equilibrium $F_t = F_c$:

$$E_c(e_c cd/2) = E_t[e'_t(1 - c)d]/2$$

Cancelling terms and rearranging:

$$e_c c = [E_t/E_c][e'_t(1 - c)]$$

Substituting Equation 4.14 for e'_t:

$$c = E_t/E_c[(1 - c)^2/c]$$

Rearranging terms:
$$c^2(1 - E_t/E_c) + 2c(E_t/E_c) - E_t/E_c = 0 \qquad (4.16)$$

The quadratic equation for c is evaluated based on the ratio of tensile and compressive MOE. Defining R equal to E_t/E_c, Figure 4.6 is a plot of the relationship between the ratio R and the constant K, used to calculate the estimated ultimate tensile stress from the MOR test. Equation 4.13 is reduced to one constant term K:

$$f'_t = (1 - c)(3PL/4bd^2) = K(PL/bd^2) \qquad (4.17)$$

These results show that for a ratio of the tensile and compressive MOE ranging from 0.25 to 1.0, a corresponding variation in the constant term K varies from about 1.1 to 1.5.

These calculations used to estimate the ultimate tensile strength for the MOR test assume that the ratio of the tensile and

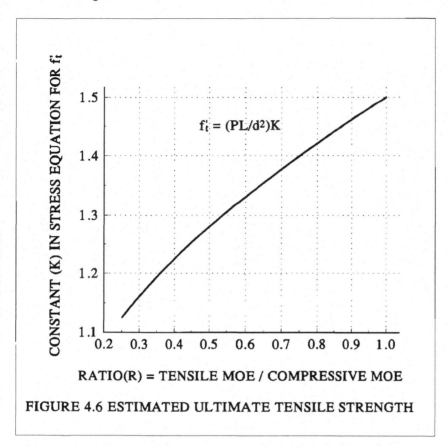

FIGURE 4.6 ESTIMATED ULTIMATE TENSILE STRENGTH

compressive MOE is known from material test on the subject refractory and that the tensile MOE is less than the compressive MOE. It is also assumed, in the above calculations, that the ratio of tensile MOE to the compressive MOE, as illustrated in Figure 4.5 is constant. That is, regardless of the stress magnitude, the ratio R does not change.

Other mathematical approaches are used to evaluate the impact of the difference in MOE as a function of the stress sign. The equivalent area method [6] is an alternate approach to account for the difference in MOE values. The equivalent area method is used to reinforce concrete design to account for the presence of steel reinforcement within the concrete.

D. MOR Test Method (Stress Sign- and Stress
Magnitude-Dependent MOE)

In some cases the stress-strain data define the stress-strain relationships as being non-linear (see Figure 4.7). As a result the MOE is dependent not only on the sign of stress, but also on the magnitude of stress.

As the definition of the stress-strain behavior becomes more complex, as shown in Figure 4.7, simple algebraic geometric relationships cannot be used. The strain still remains linear through the beam section as shown in Figure 4.4. Plane sections remain plane. However, the stress is non-linear through the beam cross section as reflected by the non-linear stress-strain relationship. The solution to calculating the estimated ultimate tensile stress now becomes more complicated and requires a higher level of analytical skills.

For non-linear stress-strain data that can be mathematically defined, classical mathematics can be used to evaluate the MOR test results in determining the estimated ultimate tensile stress. Since stress is non-linear with respect to strain, the center of gravity of the stress volumes are not identical to the strain volumes. That is, because the stress is non-linear with respect to strain, the stress volumes are not triangular in shape as are the strain volumes. As a result, the distance (a) between the compressive and tensile forces (F_c and F_t) is no longer at the center of gravity of the corresponding compressive and tensile triangular-shaped strain volumes. The reason is the geometry of the stress volumes is not reflected by the geometry of the strain volumes.

The estimated ultimate tensile strength, using classical mathematics can be determined starting with the evaluation of the internal compressive and tensile forces. Returning to Figure 4.4, the compressive and tensile bending strain through the beam midlength cross section is defined in the x direction. The vertical direction through the beam midlength cross section is defined as the y direction.

The linear relationship for the compressive bending strain

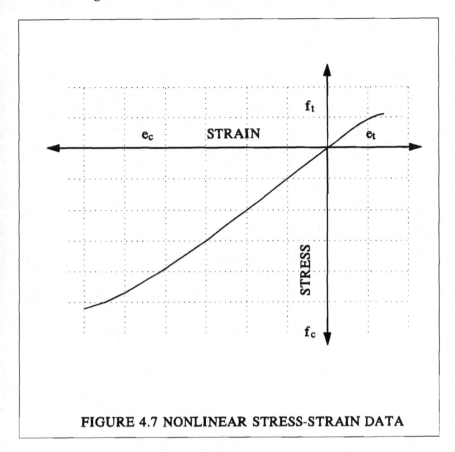

FIGURE 4.7 NONLINEAR STRESS-STRAIN DATA

through the beam cross section is:

$$e_{xc}/e_c = y/cd$$

or:

$$e_{xc} = e_c y/cd \tag{4.18}$$

where e_{xc} is the compressive bending strain in the x direction at an arbitrary distance y from the neutral axis. e_c is the maximum compressive bending strain when e'_t is achieved. e_{xt} is the tensile bending strain in the x direction at an arbitrary distance y from the neutral axis.

Likewise, for the tensile strain:

$$e_{xt} = e'_t y/[(1 - c)d] \qquad (4.19)$$

Equation 4.19 shows the linear relationship for the x-direction strain in terms of the estimated ultimate compressive and tensile strains.

The compressive stress-strain relationship plotted in Figure 4.7 is assumed to be an nth order polynomial function defined as:

$$f_c = F_c(e_c) = A_c e_c^n + B_c e_c^{h-1} + - - - \qquad (4.20)$$

Likewise, the tensile stress-strain relationship is:

$$f_t = F_t(e_t) = Ate_t^n + Bte_t^{n-1} + - - - \qquad (4.21)$$

The Equations 4.18 and 4.19 are substituted into Equations 4.20 and 4.21, respectively. The compressive and tensile x-direction bending stress at the distance y from the neutral axis can be defined as:

$$f_{xc} = A_c[e_c y/cd]^n + B_c[e_c y/cd]^{n-1} + - - - \qquad (4.22)$$

$$f_{xt} = A_t[e'_t y/(1 - c)d]^n + Bt[e'_t y/(1 - c)d]^{n-1} + - - - \qquad (4.23)$$

The internal forces F_c and F_t can be evaluated by the two integrals (again assuming rectangular cross sections):

$$F_t = b \int f_{xt} \, dy \qquad (4.24)$$

$$F_c = b \int f_{xc} \, dy \qquad (4.25)$$

After the two integrals are evaluated, the relationship between e'_t and e_c, as defined by Equation 4.14, still holds. Equation 4.24 can be defined in terms of e'_t. Therefore, the value of c can be determined, thus locating the neutral axis.

The next objective is to locate the center of gravity (or moment arms) of the internal compressive force and internal tensile force. Therefore, moment equilibrium is:

$$M_{EXT} = M_{INT} = F_c a_c + F_t a_t \tag{4.26}$$

where a_c and a_t are the moment arms of the internal compressive force and internal tension force, respectively. The compressive portion of the internal moment is:

$$F_c a_c = b \int f_{xc} \, y \, dy \tag{4.27}$$

Likewise, the tensile portion of the internal moment is:

$$F_t a_t = b \int f_{xt} \, y \, dy \tag{4.28}$$

Stress (f) is a function of strain (e), $f = f(e)$, as illustrated in Figure 4.7. And since the elastic modulus is the slope of the stress-strain curve, the compressive portion of the internal moment can be expressed as:

$$F_c a_c = b \int [df(e_c)/de_c] e_c \, y \, dy \tag{4.29}$$

$$F_t a_t = b \int [df(e_t)/de_t] e_t \, y \, dy \tag{4.30}$$

The substitution of Equation 4.14 can still be used in Equations 4.29 and 4.30. Once the moment arms a_c and a_t are determined, the internal moment can be defined by summation of moments at the location of the compressive force (as was done in Equation 4.10), that is:

$$F_t(a_c + a_t) = PL/4$$

The tensile force is expressed in terms of the estimated ultimate tensile stress and the tensile stress volume (as was done in Equation 4.12). Therefore, the estimated ultimate tensile stress, f_t, is evaluated as a result of the non-linear stress-strain relationships from the tensile and compressive stress-strain conditions.

E. STS Test Method

The STS test method is basically a two-dimensional test method in which the two stress states, f_y and f_x, differ in sign. The maximum compressive stress (f_y), located at the center of the disk specimen, is defined as:

$$f_y = -6P/\pi db \qquad (4.31)$$

where P is the total applied load, d is the diameter of the disc, and b is the thickness of the disc.

The maximum tensile stress (f_x), located at the center of the disc, is defined as:

$$f_x = 2P/\pi db \qquad (4.32)$$

Because of the two-dimensional stress state, there is a cross coupling of the two stress states. The cross coupling of the two stress states is accomplished by the mechanical material property defined as Poisson's ratio (μ). The tensile strain at the center of the disc (for a linear MOE) is defined as:

$$\epsilon_{xt} = (f_x - \mu f_y)/MOE \qquad (4.33)$$

Solving for f_x:

$$f_x = \epsilon_x MOE + \mu f_y \qquad (4.34)$$

or:

$$f_x = \epsilon_x MOE + \mu \epsilon_y MOE \qquad (4.35)$$

Using the stress sign dependent MOE:

$$f_x = \epsilon_x MOE_t + \mu \epsilon_y MOE_c \qquad (4.36)$$

Since the y strain (ϵ_y) is negative, the x stress (f_x) is lessened. However, this occurs regardless of the stress dependency of the MOE. Typically, the MOE is less for the tensile stress state, which means a lower f_x tensile stress. As a result, it takes a greater load to reach the true ultimate tensile strain. In the ASTM tests, a linear MOE is assumed, resulting in the incorrect interpretation that a greater ultimate tensile strength is reached in the STS test than the ultimate tensile strength measured in the DTP test.

For those who desire a more complete treatise on the mathematical derivation of the STS stress-strain equations, Jaeger and Cook [6] provide a complete presentation.

F. Summary

The stress sign-dependent MOE has two distinct influences on the results of the MOR test and the STS test. With the DTP test, the stress sign-dependent MOE has no influence. In the DTP, the magnitude of the tensile strain is not altered regardless of the linearity of the MOE.

With respect to the MOR test, the stress sign-dependent MOE changes the regions of compressive strain and tensile strain within the test sample. The MOR beam specimen has a one-dimensional stress state. That is, the internal bending strength of the beam specimen is a one-dimensional stress running parallel to the beam length. With lower values of tensile MOE, two changes occur. First, the volume of the tensile strain regions are increased because of the upward shift of the neutral axis. Secondly, the tensile strain becomes less than the compressive strain.

With respect to the STS test, the regions of tensile and compressive strains basically remain unchanged regardless of the non-linearity of the MOE. The STS disc specimen has a two-dimensional stress state. One stress component within the disc is the compressive stress, parallel to the external load direction, that equilibrates the external load. The second stress component is the tensile stress perpendicular to the external load. This second stress component is developed due to the cross coupling of the stress components as defined by Poisson's ratio. However the cross coupling causes a reduction in the tensile strain, as governed by the value of Poisson's ratio and lower tensile stress when a lower tensile MOE exists.

IV. ANALYTICAL INVESTIGATIONS OF ASTM
TENSILE STRENGTH TESTS USING
FINITE ELEMENT ANALYSIS

A. Introduction

As shown, by using classical mathematics, the MOR and the STS tests predict different ultimate tensile strength than the DTP test when the MOE is dependent on the sign of the stress. In order to provide additional understanding on the influence of stress sign dependent MOE on the MOR and DTP test results, the finite element analytical method is used to duplicate the stress-strain states that exist in the DPT and STS test specimens.

The ANSYS Finite Element Program [13] is used to conduct the analysis. For those who desire additional reading on the finite element method (FEM) and available FEM computer programs review references 14 through 20.

The following discussion addresses the finite element analysis of the MOR test specimen stress state and the STS test specimen stress state. As previously discussed, the DTP test specimen is a one dimensional tensile stress-strain state. There is no compressive stress-strain condition that interferes with the tensile stress-strain condition. Therefore, the DTP test is not evaluated.

B. MOR Test by FEA

A finite element analysis was conducted on the MOR test specimen using both the linear MOE and a stress sign-dependent MOE. The model used in both cases was identical. The MOR test specimen geometry was assumed to be a beam 152 mm (6 in.) long and 25 mm by 25 mm (1 in. x 1 in.) in cross section. The two point support was used here with the support being at 127 mm (5 in.) center to center. Because of both load and geometric symmetry only half of the beam geometry was required (see Figure 4.8).

The analysis was conducted assuming that the DPT test resulted in an ultimate tensile strength of 1.38 MPa (200 psi) for the hypothetical material used in the following example. From

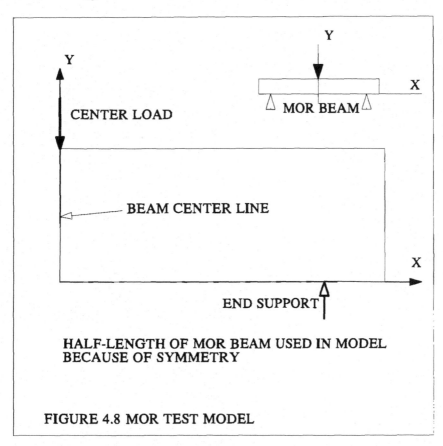

FIGURE 4.8 MOR TEST MODEL

Equation 4.3, a load (P) of 119 N on the MOR specimen was required to create a maximum tensile stress of 1.38 MPa. The beam model had a rather fine mesh of quadrilateral elements (two-dimensional linear strain within quadrilateral) with 40 elements in the beam depth and 60 elements along the beam half-length.

The FEA-predicted stress results for the MOR beam specimen with a linear MOE are described in Figure 4.9. The MOE throughout the beam was assumed to be 2×10^4 MPa (3×10^6 psi) regardless of stress state. As shown, the zero bending stress (C contour line) is at the midheight neutral axis for the full beam length between the support points. The maximum tensile stress at

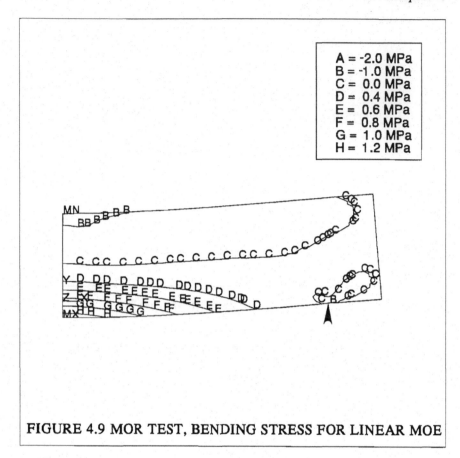

FIGURE 4.9 MOR TEST, BENDING STRESS FOR LINEAR MOE

the bottom, midlength position on the beam specimen is 1.35 MPa (192 psi), close to the predicted 1.38 MPa ultimate tensile strength from the DPT test. See Table III for a summary of results. The primary tensile stress-strain region of the beam specimen is defined as extending vertically from the bottom of the beam to the midheight neutral axis (tick mark at N.A.), and extending horizontally from support to support.

Figure 4.10 describes the computed region of tensile bending stress (dark region). Note the tensile stress bulbs in the regions of the two support points. Except for the region surrounding the supports, the tensile bending stress region can be approximated as a rectangular area. However, the tensile stress is quite low in these

Table III

Comparison of Predicted Tensile
Stress for MOR and STS Tests

Maximum Tensile Stress by Test Type, MPa (PSI)				
Method	MOR		STS	
	Linear MOE	Stress Sign MOE*	Linear MOE	StressSign MOE*
Finite Element Method	1.32 (192)	1.03 (149)	1.37 (199)	1.29 (187)
Classical Equation	1.38+ (200)+		1.38++ (200)++	
	1.27+++ (184)+++		1.17* (170)*	
			1.59** (231)**	

+   Equation 4.3 for MOR test.
++   Equation 4.32 for STS test.
+++ Equation 4.37, Timoshenko Equation.
*   Equation 4.38 for rectangular prism.
**   Equation 4.40 for square prism.

local stress bulbs at the support regions. Therefore, the stress-strain energy is also quite small in these regions, and assuming this region as part of the rectangular area would not significantly influence the solution.

A second analysis was conducted using tensile MOE of 0.69 x 10^4 MPa (1 x 10^6 psi) and a compressive MOE of 2 x 10^4 MPa. By trial and error, repeated solutions were made until a converged stable solution of the tensile stress-strain region of the beam was achieved. The converged solution of the tensile stress-strain region is described by the lightly shaded area in Figure 4.11. This region is an approximate area of the primary tensile region.

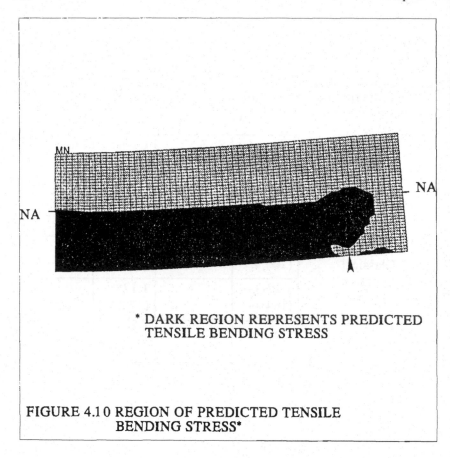

* DARK REGION REPRESENTS PREDICTED
  TENSILE BENDING STRESS

FIGURE 4.10 REGION OF PREDICTED TENSILE
BENDING STRESS*

The region of calculated tensile bending stress is described in
Figure 4.12 (dark region).   Compared to the assumed tensile
bending stress region shown in Figure 4.11, the calculated region
provides a reasonable duplication of the assumed region. The tick
marks at the ends of the MOR specimen model identify the top
limit of the assumed tensile bending stress region. The tick mark to
the left of the support point identifies the right limit of the assumed
tensile bending stress region (region of tensile MOE). The assumed
region is a bit larger in the region near the support. However, the
difference in these two regions (calculated tensile stress and tensile
MOE) represents the portion of low strains and, therefore, low
strain energy. As a result, the solution results would not be altered
significantly if the true tensile bending stress region was incorporated

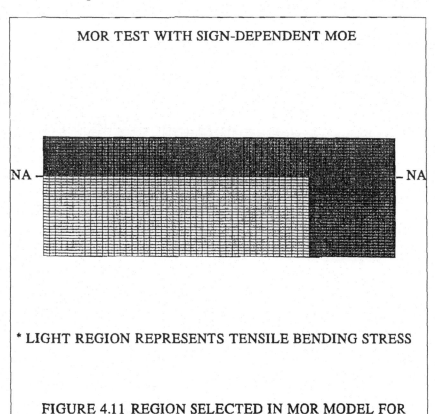

MOR TEST WITH SIGN-DEPENDENT MOE

NA — — NA

* LIGHT REGION REPRESENTS TENSILE BENDING STRESS

FIGURE 4.11 REGION SELECTED IN MOR MODEL FOR
THE TENSILE* MOE

into the MOR specimen model.

As in the case of the linear MOE, the support point causes a tensile stress bulb just above the support. This is assumed to be of little consequence in the solution because of the low stress and strain in this region. It should also be remembered that the beam Equation 4.3 assumes these support point secondary stresses and strains have no influence on the solution. This assumption is verified in the MOR analysis with the linear MOE.

In Figure 4.13, the stress contour plot for the stress sign-dependent MOE, the neutral axis shifts upward to a position of

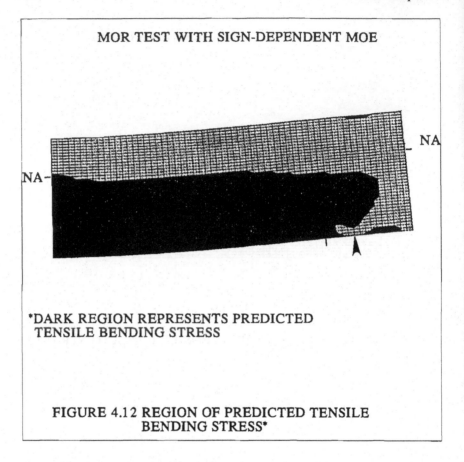

MOR TEST WITH SIGN-DEPENDENT MOE

*DARK REGION REPRESENTS PREDICTED
TENSILE BENDING STRESS

FIGURE 4.12 REGION OF PREDICTED TENSILE
BENDING STRESS*

about 17 mm measured from the bottom of the beam. Note that the
beam mid-height is 12.5 mm as measured from the bottom or top of
the beam. The maximum tensile stress is 1.03 MPa (149 psi) for the
same loading used with linear MOE analysis.

The comparison of the two MOR analyses is summarized in
Table III. The lower valued tensile MOE results in a lower
maximum tensile stress. The assumption that the tensile MOE and
compressive MOE are constant with respect to the stress magnitude,
as illustrated in Figure 4.5, is most likely not realistic. In actuality,
the tensile MOE decreases with increasing tensile stress as shown in
Figure 4.7.

BEAM BENDING STRESS

A =	-2.0 MPa
B =	-1.0 MPa
C =	0.0 MPa
D =	0.2 MPa
E =	0.4 MPa
F =	0.6 MPa
G =	0.8 MPa
H =	1.0 MPa

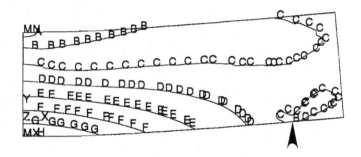

FIGURE 4.13 MOR TEST, BENDING STRESS FOR STRESS
SIGN-DEPENDENT MOE

C. STS Test by FEA

Similar analyses were conducted on the STS test specimen. Because of load and geometric symmetry, only a quarter section of the disc was required, as described in Figure 4.14. The y direction is vertical or parallel to the applied load. The x direction is normal to the y direction. The disc was assumed to have a 254-mm (10-in.) diameter and a 25-mm (1-in.) thickness. Based on Equation 4.32, a 14,000-N (3,140-lb.) load was applied to create an f_x tensile stress of 138 MPa (200 psi) at the center of the disc. The resulting x and y stresses are described in Figures 4.15 and 4.16, respectively, for the case of a linear MOE. Note that the vertical y stress is compressive through the full area of the disc. The

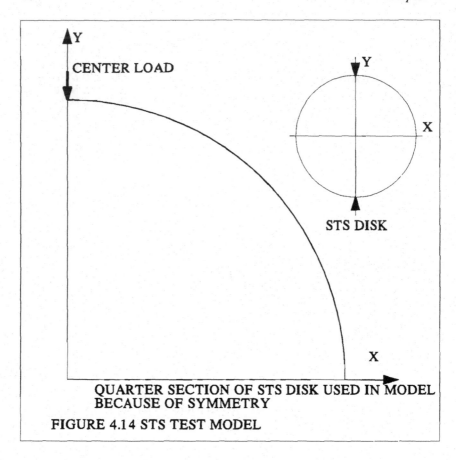

QUARTER SECTION OF STS DISK USED IN MODEL
BECAUSE OF SYMMETRY

FIGURE 4.14 STS TEST MODEL

horizontal x stress is tensile throughout the major portion of the disc. A locally high compressive x stress exists in the region of the load. However, the major portion of the disc has a tensile x stress.

It should also be noted that the x stress is not uniform across the height of the disc as defined in several investigations using classical mathematics. According to the finite element solution, the tensile x stress increases along the vertical centerline from zero values near the load points to maximum values at the center. Only near a small center region of the disc does the tensile x stress become nearly constant.

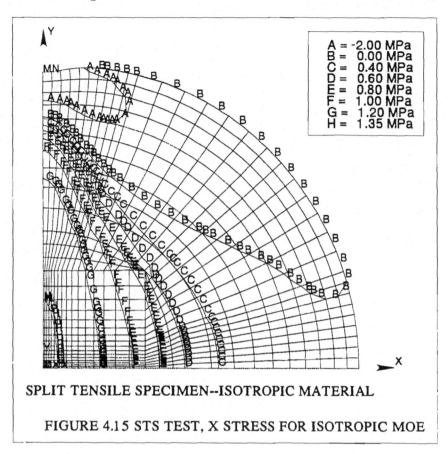

SPLIT TENSILE SPECIMEN--ISOTROPIC MATERIAL

FIGURE 4.15 STS TEST, X STRESS FOR ISOTROPIC MOE

The FEA results of the STS test with the stress sign dependent MOE are summarized in Figures 4.17 and 4.18. The values of tensile and compressive MOE are identical to the values used in the MOR finite element investigation. However, in the STS disc the two-dimensional stress state requires using an orthotropic definition of the material properties. Therefore, the tensile MOE (in the x direction) was 0.69×10^3 MPa while the compressive MOE (in the y direction) was 2×10^4 MPa.

The maximum tensile x stress at the center of the disc for the stress sign-dependent MOE is about 1.29 MPa (187 psi). This predicted tensile stress is slightly less in value to the linear MOR results.

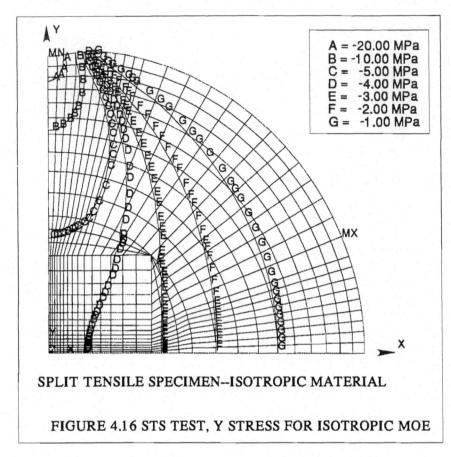

A = -20.00 MPa
B = -10.00 MPa
C = -5.00 MPa
D = -4.00 MPa
E = -3.00 MPa
F = -2.00 MPa
G = -1.00 MPa

SPLIT TENSILE SPECIMEN--ISOTROPIC MATERIAL

FIGURE 4.16 STS TEST, Y STRESS FOR ISOTROPIC MOE

D. Summary

The finite element investigations of the MOR and STS tests assist in providing an understanding with regard to the impact of the MOE on the predicted stresses. Most brittle materials have a tensile MOE of lesser value than the compressive MOE. The result being that the MOR and STS tests develop a lower tensile stress than tensile stress developed in the DTP test. As a result, greater loads are imposed on the MOR and STS specimens during testing to cause fracture of the specimens at the true tensile strength. When the classical equations are employed for a linear MOE material, the MOR and STS tests predicted ultimate tensile stress to be both lesser and greater than the DTP test.

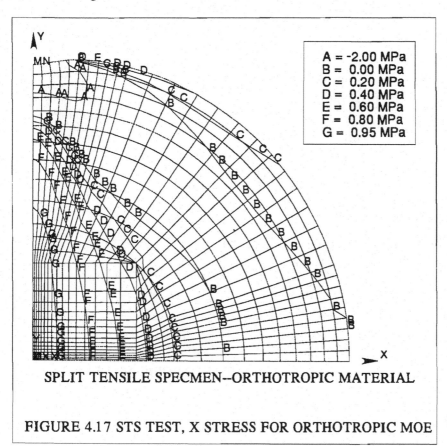

SPLIT TENSILE SPECMEN--ORTHOTROPIC MATERIAL

FIGURE 4.17 STS TEST, X STRESS FOR ORTHOTROPIC MOE

There are other causes for the differences in the predicted ultimate tensile stress. Most of these other causes are attributed to misalignment of the specimen in the testing rig, misalignment of the testing rig or imperfections in the specimen geometry.

The MOR test predicted ultimate tensile stress is calculated from Equation 4.3 as:

$$f_t = 1.5 \ PL/bd^2 = (1.5 \ P/d)(L/d)$$

Duckworth [1] cites the Timoshenko [21] equation for calculating the beam tensile bending stress, which is modified to account for the localized stresses that occur due to the point load on

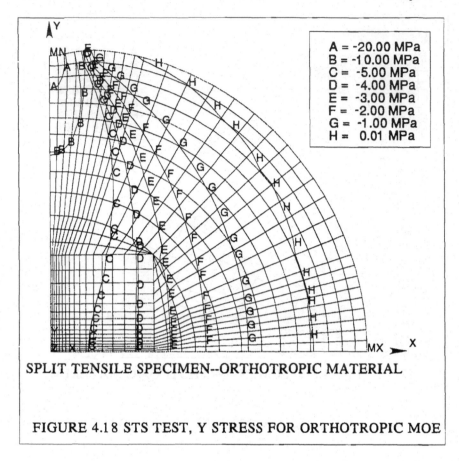

SPLIT TENSILE SPECIMEN--ORTHOTROPIC MATERIAL

FIGURE 4.18 STS TEST, Y STRESS FOR ORTHOTROPIC MOE

the beam. The Timoshenko equation is defined as:

$$f_t = (1.5P/bd)(L/d - 4/3\pi) \tag{4.37}$$

According to this equation, the tensile bending stress is reduced by subtracting $4/3\pi$ from L/d. Substituting the values from the Section III.B MOR test with linear MOE (P = 119 N, d = 25.4 mm, b = 25.4 mm and L = 127 mm) into Equation 4.37:

$$f_t = (1.5 \times 119/25.4^2)(127/25.4 - 4/3\pi)$$

$$= 0.277 (5 - 0.42) = 1.27 \text{ MPa (184 psi)}$$

The comparison of the predicted tensile bending stress from the classical beam bending stress equation, the Timoshenko modified beam bending stress equation and MOR test results indicates that the Timoshenko modification is too conservative. That is, the effects of the concentrated load, in the region of maximum tensile bending stress, is not as great as defined by the parameter $4/3\pi$.

V. FINITE ELEMENT ANALYSIS OF OTHER TYPES OF DIAMETRAL LOADED TENSILE STRENGTH TEST SPECIMENS

A. Introduction

Other types of specimen geometries [6] have been used to evaluate the tensile strength of brittle materials. The tensile stress prediction by these other types of specimens is the focus of the following discussions.

Tensile strength of brittle materials can be evaluated using other types of specimen geometries. These other geometries are selected either for the purpose of reducing the cost of specimen preparation or for the purpose of obtaining a more accurate determination of the tensile strength. The following analyses and discussions address these additional types of diametral compression tests using test specimen geometries defined as the rectangular prism, the square prism, the sphere, and the hollow cylinder (or ring). The following analyses are limited to a linear MOE.

B. Rectangular Prism Analysis

The rectangular prism analysis was conducted on a rectangular specimen 254 mm (10 in.) in depth (d) and 500 mm (20 in.) in length (L). Diametrical compressive loads were applied just with the STS test specimen. Since the STS test specimen had a diameter of 254 mm, the predicted tensile stress results of these two specimens could be compared.

The maximum tensile stress at the center of the rectangular prism is defined [6] as:

$$f'_t = 1.70 \ P/\pi db \qquad\qquad (4.38)$$

where d, in this case, is the depth of the rectangular specimen. Comparing Equations 4.32 and 4.38, the disc specimen should develop a slightly greater tensile stress.

Using a diametrical load P of 14,000 N (3140 lb.), the estimated ultimate tensile stress (using Equation 4.38) for the rectangular prism test specimen is:

$$f'_t = 1.7(14,000)/\pi(254)(25.4) = 1.17 \ \text{MPa (170 psi)}$$

The finite element analysis was conducted using a quarter section model as described in Figure 4.19. Because of both load and geometric symmetry, only a quarter section of the rectangular prism specimen was required to model the specimen.

Of particular interest was the magnitude of the horizontal stress (x stress direction) and the vertical stress (y stress direction). Figures 4.20 and 4.21 are computer plots describing the x and y stress direction, respectively, in the quarter section model. Just as in the previous finite analysis, MX, is the location of maximum tensile stress and MN maximum compressive stress (or minimum stress as defined on an algebraic scale).

With respect to the x stress described in Figure 4.20, the maximum tensile and compressive x stress is located in the region of the applied load. Because of the density of the stress contours, the MX and MN are difficult to distinguish. Figure 4.22 is an amplification of the region in the vicinity of the load (including the element grids). The depressed area of the applied load describing the x stress is also the location of the maximum compressive x stress (MN). Just adjacent to the applied load is the location of the maximum tensile x stress (MX). However, the maximum tensile x stress is not in a region of a continuous field of tensile stress. Therefore, fracturing or cracking would be limited in this region. Fracturing also initiates in the center of the specimen, midway

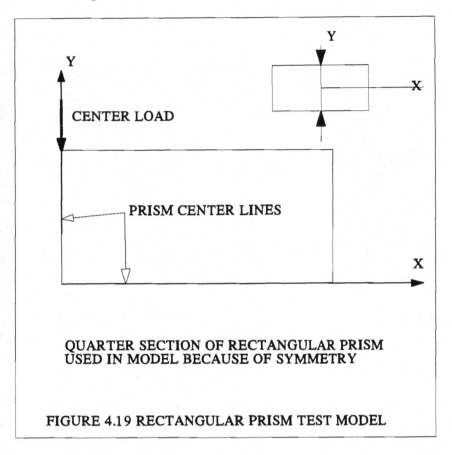

Y

Y

X

CENTER LOAD

X

PRISM CENTER LINES

**QUARTER SECTION OF RECTANGULAR PRISM
USED IN MODEL BECAUSE OF SYMMETRY**

FIGURE 4.19 RECTANGULAR PRISM TEST MODEL

between the two opposing loads because of the maximum tensile stress in this region. And, in addition, because the tensile x stress is continuous here, the propagation of a vertical crack and the ultimate splitting of the specimen is achieved. As previously noted, Figure 4.22 also describes the fine mesh of the elements used to capture the stress gradients. The predicted maximum tensile x stress near the load was about 2.13 MPa (309 psi). The predicted maximum tensile x stress at the center of the specimen was 1.08 MPa. The predicted maximum tensile stress at the specimen center is slightly less than the maximum tensile stress predicted by Equation 4.38. The tensile x stress results are summarized in Table IV.

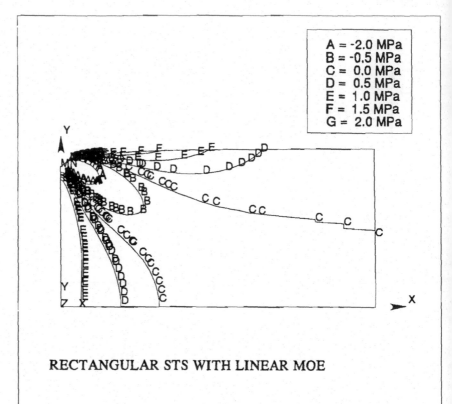

RECTANGULAR STS WITH LINEAR MOE

FIGURE 4.20 RECTANGULAR PRISM, X STRESS FOR
ISOTROPIC MOE

C. Square Prism Analysis

The square prism analysis was conducted on a square specimen 254 by 254 mm (10 by 10 in.). The same magnitude of diametrical loads as used in the STS and rectangular specimens were used for the square specimen.

The tensile stress at the center of the square prism to cause splitting is defined by Jaeger [6] as:

$$f'_t = 0.366 \ P/R \tag{4.39}$$

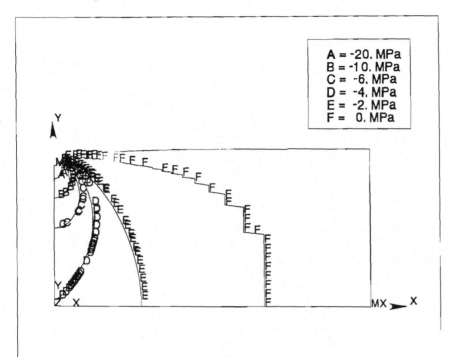

RECTANGULAR STS WITH LINEAR MOE

FIGURE 4.21 RECTANGULAR PRISM, Y STRESS FOR
ISOTROPIC MOE

where R is the half-length of the sides of the square prism. According to Jaeger, Equation 4.34 was developed for a thickness-to-side length ratio of 0.125. However, it was confirmed in the following analysis that the thickness-to-side length ratio has no relevance to constant values in Equation 4.39. Equation 4.39 can be rewritten in the form of the previous equations in which d equals 2R. Substituting d/2 for R:

$$f'_t = 0.732 \ P/d$$

Using the constant π, the equation takes the form:

$$f'_t = 2.3 \ P/\pi db \qquad (4.40)$$

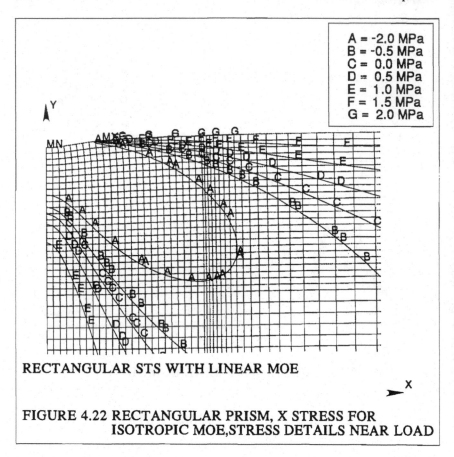

RECTANGULAR STS WITH LINEAR MOE

FIGURE 4.22 RECTANGULAR PRISM, X STRESS FOR
                    ISOTROPIC MOE, STRESS DETAILS NEAR LOAD

The Equation 4.39 (or Equation 4.40) predicts a higher tensile stress
for the square specimen than that predicted by Equation 4.32 for the
split disc specimen of identical size.

The x and y stress contours for the square prism are
summarized in Figures 4.23 and 4.24, respectively. Comparing
Figures 4.23 and 4.24 with the corresponding STS results (Figures
4.15 and 4.16) shows that the square prism stress contours agree
favorably with the STS stress contours. Also, the maximum tensile
stresses of the square prism are about identical to maximum tensile
stresses of the STS. It can be concluded from these studies that the
square prism tensile stresses, for all intents and purposes, identical
to the STS tensile stresses. That is, Equation 4.40 does not

Table IV

Comparison of Predicted Tensile Stress
By Various Diametrically Load Specimens

Method	Type of Diametrical Loaded Specimen; Predicted Tensile Stress, MPa (PSI)		
	Rectangular Prism Specimen	Square Prism Specimen	Spherical Specimen
Finite Element Method	1.08# (159) 2.13## (309)	1.36# (197) 1.72## (250)	1.5# (220) 11.7## (1670)
Classical Equation	1.17+ (170)	1.38++ (200) 1.59* (230)	0.38 to 0.75+++ (55 to 109)

#	Maximum tensile stress at interior of model
##	Maximum tensile stress in vacinity of load
+	Equation 4.38 for rectangular prism
++	Equation 4.32 for STS test
+++	Equation 4.42 for spherical prism
*	Equation 4.40 for square prism

realistically predict the square prism subject tensile stresses. The more accurate Equation 4.32 can be used for either the STS test or the square prism test.

Details of the x stress are shown in Figure 4.25. Again the square prism x stress is very similar to the STS x stress described in Figure 4.15. However, the horizontal top boundary of the square prism creates x stress that is not present in the STS. The curved boundary of the STS results in no similar x stress. However, the square prism boundary x stresses do not affect the x stress values at the center of the square prism.

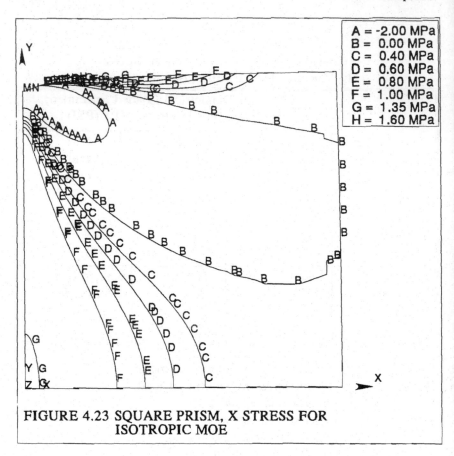

FIGURE 4.23 SQUARE PRISM, X STRESS FOR
            ISOTROPIC MOE

D. Spherical Specimen Analysis

The spherical specimen has been suggested [6] as another
geometry for evaluating the tensile strength of brittle materials. To
be consistent with the previous specimen's size, the spherical
specimen analyzed was assumed to be 254 mm in diameter.

The equation used to define the maximum tensile stress that
would cause the specimen to split is defined [6,22,23] as:

$$f'_t = k \, P/R^2 \tag{4.41}$$

or in terms of the specimen diameter:

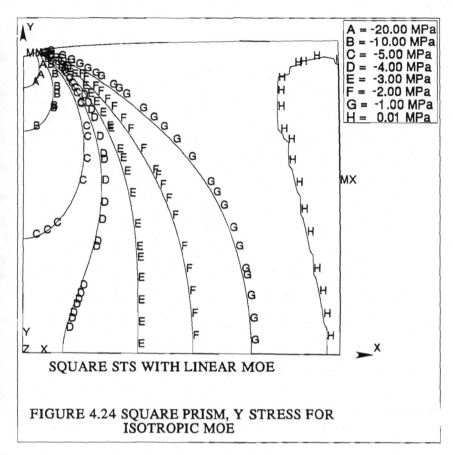

SQUARE STS WITH LINEAR MOE

FIGURE 4.24 SQUARE PRISM, Y STRESS FOR
ISOTROPIC MOE

$$f'_t = 4k \ P/d^2 \qquad (4.42)$$

According to Jaeger [6], the value of k varies from 0.225 to 0.45
based on the investigator [24-26].

The spherical specimen model is an axisymmetric quarter
section of the sphere. This is the only portion of the geometry that
is required due to symmetry of both geometry and loading.
Diametrical loading 28,000 N (6,280 lb.) was applied to the model.

Figure 4.26 describes the radial stress (stress in radial x
direction) within the spherical specimen. The radial stress pattern

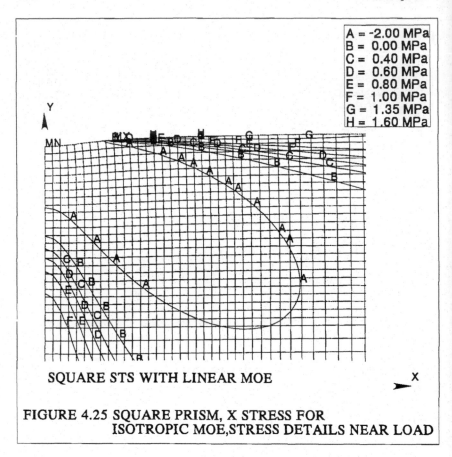

SQUARE STS WITH LINEAR MOE

FIGURE 4.25 SQUARE PRISM, X STRESS FOR
                    ISOTROPIC MOE,STRESS DETAILS NEAR LOAD

differs from the previous prism geometry results in that the
maximum stress occurs at an internal region near the load. In the
previous specimens, the maximum stress occurs at the midregion of
the specimen.   The maximum tensile stress to cause fracture is
about 1.5 MPa (220 psi).

A greater tensile stress of about 11.7 MPa (1670 psi) exists at
the surface adjacent to the load (see Figure 4.27). Figure 4.27 is an
amplification of the region surrounding the load.   The depressed
area identifies the region of the applied diametral load. It should be
explained   that   localized   cracking   will occur in the tensile
region adjacent to the loading. However, this tensile region is quite
limited and the tensile fracturing would be quickly arrested when

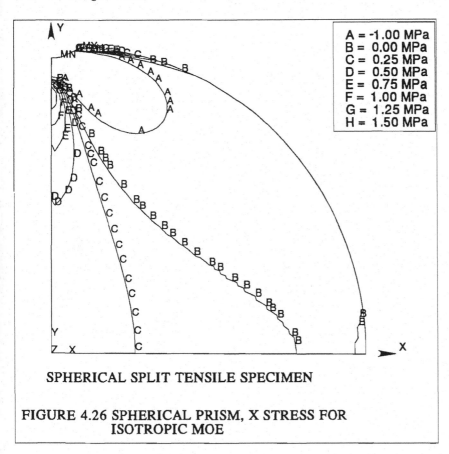

SPHERICAL SPLIT TENSILE SPECIMEN

FIGURE 4.26 SPHERICAL PRISM, X STRESS FOR
            ISOTROPIC MOE

the crack tip reaches the compressive stress region. The C contour
is the zero stress contour and identifies the line of transition from
the tensile stress to compressive stress. As a result this localized
tensile stress region will not cause the spherical specimen to
fracture.

The vertical stress (y direction of sphere) is basically all
compressive as evidenced in Figure 4.28.

The circumferential stress within the sphere is described in
Figure 4.29. The maximum circumferential tensile stress occurs in
the same region as the maximum radial tensile stress. It is
concluded that, because of symmetry, the splitting or fracturing of

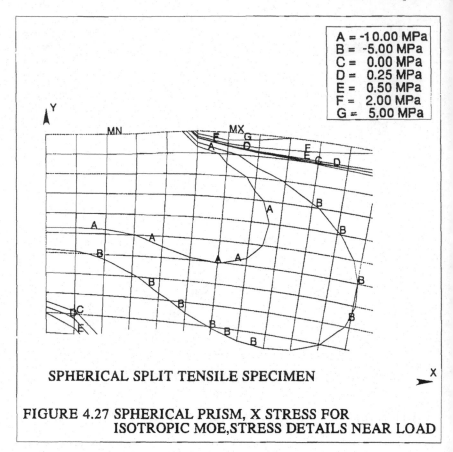

SPHERICAL SPLIT TENSILE SPECIMEN

FIGURE 4.27 SPHERICAL PRISM, X STRESS FOR
              ISOTROPIC MOE, STRESS DETAILS NEAR LOAD

the spherical specimen initiates simultaneously at two regions
within the specimen near the load points.

Substituting the results from the analysis, the predicted value
for k in Equations 4.41 or 4.42 is 0.875. This value for k is about
2 to 4 times greater than the value defined by previous
investigators.

Of all the prism geometries, the spherical prism appears to
have the greatest disparity between actual material test results and
analytical predictions of tensile strength. The spherical prism also
has the greatest disparity in test results as defined by
previously referenced investigators.

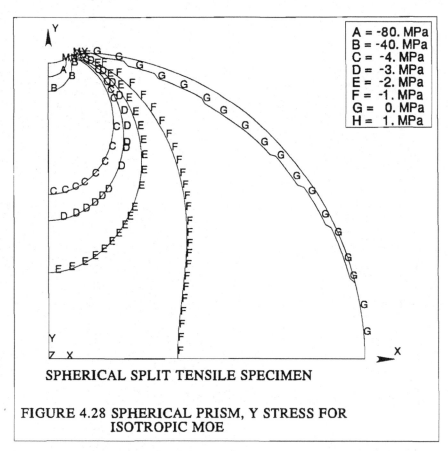

SPHERICAL SPLIT TENSILE SPECIMEN

FIGURE 4.28 SPHERICAL PRISM, Y STRESS FOR
ISOTROPIC MOE

The observed variability of tensile strengths in using the spherical prism is perhaps explained by three stress-strain features of the spherical prism that make it different from the previously defined prisms used for predicting tensile strength. First, the spherical prism is a three dimensional prism. All of the other tensile strength prisms were either one-dimensional or two-dimensional with regard to the stress-strain states created within the specimen. The second unique stress-strain feature of the spherical prism is the biaxial tensile stress-strain state at the two interior regions identified in Figures 4.26 and 4.29. At these two regions peak radial and circumferential tensile stress states exist simultaneously. The third unique stress-strain feature of the spherical prism is that two equal peak tensile stress-strain states

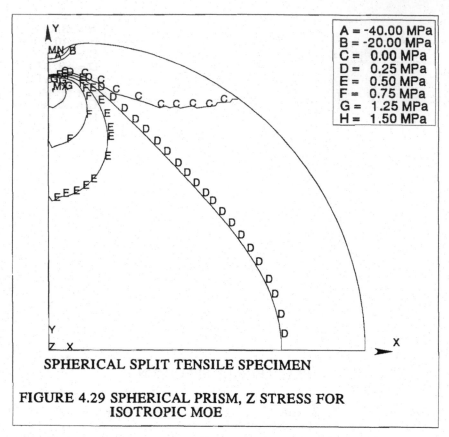

SPHERICAL SPLIT TENSILE SPECIMEN

FIGURE 4.29 SPHERICAL PRISM, Z STRESS FOR
             ISOTROPIC MOE

exist simultaneously. In all of the other tensile strength specimens, only one peak tensile stress-strain state exists within the specimen geometry.

VI. SUMMARY

The various tensile strength geometries provide a range of values for estimating tensile strength for brittle materials. The primary reason for this range is attributed to the stress-dependent modulus of elasticity. The dimension of the stress state, in combination with the stress-dependent elastic modulus, results in maximum tensile stress-strain states that differ from the tensile stress predicted by the classical equations (see summary in Table III). The analytical results support the observed variations in tensile strength observed by investigators using the direct pull test, the modulus of rupture test and the splitting tensile strength test.

Other specimen geometries have been proposed to evaluate the tensile strength of brittle materials. Analytical investigations of these geometries indicate that some of the proposed equations to be used for these geometries are not accurate.

As summarized in Table IV, the classical equation for the rectangular specimen predicts a tensile strength slightly greater than the analytical results.

The classical equation used for the square specimen predicts a much higher tensile strength than that predicted by analysis. The STS classical equation more accurately predicts the tensile strength for the square specimen.

The spherical specimen has the greatest variation in predicted tensile strength as determined by the spherical specimen classical equation. The analytical results identify unique features of the stress state in the spherical prism that may account for the wide range in tensile strength tests observed by some investigators.

REFERENCES

1. Duckworth, W. H., Precise Tensile Properties of Ceramic Bodies, Journal of the American Ceramic Society, 34[1], 1-9, 1951.

2. ASTM Annual Book of Standards, Vol. 15.01, Section 15, Refractories; Carbon and Graphite Products; Activated Carbon, 1993.

3. ASTM Annual Book of Standards, Vol. 4.02, Section 5, Concrete and Aggregates, 1993.

4. Bauer, E. E., Plain Concrete, Third Edition, McGraw-Hill Book Company, Inc., New York, 1949.

5. Griffith, A. A., The Phenomena of Rupture and Flow in Solids, Trans. Roy. Soc. (London) Series A, 221[4], pp. 163-198 (1921).

6. Jaeger, J. C. and Cook, N. G. W., <u>Fundamentals of Rock Mechanics</u>, Halsted Press, London, p. 191, 1976.

7. Jaeger, J. C. and Hoskins, E. R., Stresses and Failures in Rings of Rock Loaded in Diametral Tension or Compression, Br. J. Appl. Phys., 17, pp. 685-692 (1966).

8. Schacht, C. A., Stress-Dependent Elastic Modulus and Its Effect on Determining the Tensile Strength of Refractory/Ceramic Materials, American Ceramic Society 92nd Annual Meeting, April 22-26, 1990.

9. Dieksen, B., Experiments to Determine the Heat Resistance of Ceramics, Special Report from the Intelligence Department, Air Material Command, Wright-Patterson Air Force Base, Dayton, Ohio (Circa 1950).

10. Mong, L. E., Elastic Behavior and Creep of Refractory Brick Under Tensile and Compressive Loads, Journal of the American Ceramic Society, Vol. 30, No. 3, pp. 69-78 (1947).

11. Boyd, J. E. and Folk, S. B., <u>Strength of Materials</u>, McGraw-Hill, New York, 1950.

12. Marin, J. and Sauer, J. A., <u>Strength of Materials</u>, MacMillian Co., New York, 1954.

13. ANSYS, Engineering Analysis System; Swanson Analysis Systems, Inc.; P.O. Box 65, Houston, PA 15342.

14. Bathe, K. J., <u>Finite Element Procedures in Engineering Analysis</u>, Prentice-Hall, Inc., New Jersey, 1982.

15. Desai, C. S., and Abel, J. F., <u>Introduction to the Finite Element Method</u>, Van Nostrand Co., New York, 1972.

16. Zienkiewicz, O. C., The Finite Element Method in Engineering Science, McGraw-Hill Co., London, 1977.

17. Pilkey, W., Saczalski, K. and Schaeffer, H., Structural Mechanics Computer Programs, The University Press of Virginia, 1974.

18. Kardestuncer, H., Finite Element Handbook, McGraw-Hill Book Co., 1987.

19. Baran, N. M., Finite Element Analysis on Microcomputers, McGraw-Hill Book Co., 1988.

20. Falk, H., Microcomputer Software for Mechanical Engineers, Van Nostrand Reinhold Co., 1987.

21. Timoshenko, S., Strength of Materials Part II, Advanced Theory and Problems, 2nd ed., Van Nostrand Co., New York, pp. 352-355, 1941.

22. Abramian, B. L., Aruntiunian, N. K. H., and Babloian, A. F., On the two-contact problem for an elastic sphere, Fizika Metall., 28, pp. 622-629 (1964).

23. Sternberg, E. and Rosenthal, F., The elastic sphere under concentrated loads, J. Appl. Mech., 19, pp. 413-421 (1952).

24. Berenbaum, R. and Brodie, I., The tensile strength of coal, J. Inst. Fuel, 32, pp.320-327 (1959).

25. Jaeger, J. C., Failure of rocks under tensile conditions, Int. J. Rock Mech. Min. Sci., 4, pp. 219-227 (1967).

26. Hiramatsu, Y., and Oka, Y. Determination of the tensile strength of rock by a compression test of an irregular test piece, Int. J. Rock Mech. Min. Sci., 3, pp. 89-99 (1966).

5
Choosing the Best Refractory for the Thermomechanical Application

I. INTRODUCTION

The objective of this chapter is to assist the user in providing criteria for selecting the strongest refractory with regard to the specific refractory lining application. The definition of *strongest* here means a refractory that will not fracture or at least have the least chance of fracturing when exposed to the load environment. Selecting the strongest refractory may require a user review of the refractory manufacturer's literature or user requests for additional tests to be conducted by the manufacturer that are unique to the user's process applications. In some cases, the user may conduct tests on the manufacturer's samples to assist in choosing the better refractory material. In many applications, the choice of the better refractory is based on both chemical and mechanical considerations. As previously discussed, the chemical aspects will not be addressed in this text. Of primary concern here is the mechanical material property aspects of choosing the best refractory.

II. BACKGROUND

As discussed in Chapter 4, a considerable amount of effort has been placed by the material engineers, material scientists and material testing communities in defining the ultimate tensile and compressive strengths of refractories. This is evidenced by the various ASTM tests that have been developed to quantify the ultimate strengths of materials and, in our case, refractory materials.

The ultimate tensile strength and ultimate compressive strength are typically the most available data provided by refractory manufacturers. This data is often examined by users to rank the mechanical strength in determining the strongest refractory material. The selection of strongest refractory is often a task of examining other property data as well. The selection of which mechanical property data to be used for ranking the various supplier's materials, however, should be based on, or at least be strongly dependent upon, the type of load environment to which the user's refractory system is subjected. As discussed in Chapter 2, a refractory structure is exposed to either stress-controlled or strain-controlled loadings or a combination of both.

There are a few types of refractory structures that are exposed to primarily stress controlled loads. These types of refractory structures may also be exposed to a significant temperature environment. In this case, the refractory lining is not restrained against thermal growth, the result being no significant thermal stress. In these cases, the stress-controlled loads are gravity weight loads which are usually quite small.

Most refractory lining structures are exposed to strain-controlled or thermal expansion loads. That is, the majority of refractory structures are restrained, to some degree, against full free expansion of the heated refractory. Therefore, the restrained refractory material is subjected to a portion of the total potential thermal strain.

The ratio between the thermal expansion stresses and the gravity load stresses is normally quite high. In other words, gravity load stresses typically will be in the range of 0.2 to 1.0 MPa (30 to

150 psi), while the thermal expansion stresses can be in the range of 15 to 40 MPa (2000 to 6000 psi) or higher. Based on the magnitude of the stress (or strain) within the refractory structure, the refractory structure subjected strain-controlled loads (thermal expansion forces) should demand attention in ranking and selecting the strongest refractory than for a stress-controlled loaded refractory structure.

III. DEFINING THE PROBLEM

The type of process contained within the refractory structure, the stiffness of the external steel support structure, the severity and cyclic nature of the transient process temperatures, the magnitude of the process temperature and the severity of the process chemistry are just a few items that influence the life of refractories. Some process vessels are stationary, while others expose the refractory lining to transient displacements (e.g., rotary kilns). Some vessels are transported within the plant while filled with molten metal (e.g. steelmaking ladles). The life of a refractory lining is a function of a combination of the severity of the chemistry and mechanical environments. Each time a lining is selected, the question always arises: which supplier provides the best refractory for my application?

Perhaps the best way to describe the thermomechanical criteria for selecting the strongest refractory is to provide examples which illustrate the choice process. The first example illustrates a refractory structure exposed to a stress-controlled load. The second example illustrates a refractory structure exposed to a strain-controlled load. Both of these examples are greatly simplified to demonstrate the primary criteria for selecting the strongest refractory.

Prior to the presentation of the two example problems, let us first assume that two competing refractory manufacturers are each proposing a refractory for your application. You, the user, must decide which refractory material is best for our application. Chapter 3 described compressive stress-strain data for some typical fired high-alumina (70% Al_2O_3) brick (see Figure 3.1). In the following examples, the complexity of the temperature environment on the refractory lining is also simplified in order to assist in defining the selection criteria. The refractory in our process is assumed to be at a single constant temperature. The magnitude of your process temperature is arbitrary at this point.

For the purpose of identification, one refractory material manufacturer is called manufacturer A. The second manufacturer is manufacturer B. It is further assumed that both materials were tested in a laboratory to evaluate the compressive stress-strain response of each material at our process temperature. Figure 5.1 would be the resulting compressive stress-strain curve obtained for each of the refractory materials for your process temperature.

Upon examination of the compressive stress-strain data of materials A and B, two distinguishing differences are identified. Material A has a much greater ultimate compressive strength (f'_{CA}) than that of material B (f'_{CB}). Material A, however, has a lesser ultimate compressive strain (e'_{CA}) than the ultimate compressive strain of material B (e'_{CB}). It can be concluded that material A is a more stiff material than material B. Likewise, material B is a softer or more flexible material than material A. Of primary interest here is the magnitude of the ultimate compressive strength. As indicated in Figure 5.1 and as reported by manufacturer A, refractory material A has a greater ultimate compressive stress than material B. Therefore, based on the test data, initial conclusions are that refractory material A is stronger than refractory material B. This may or may not be true based on our application environment.

A. Choosing the Strongest Refractory for a Refractory System Subjected to Stress-Controlled Loading

Stress-controlled loading was first introduced in Chapter 2. The refractory lining stress in the following simplified example (see Chapter 2, Figure 2.1) is defined by dividing the cross-sectional area (A) of the sample into the external loading (P). The external load is a stress-controlled load, defined here is a gravity load, imposed on the simplified geometry (refractory cube). In our case, the load P could be a combination of other external gravity loads

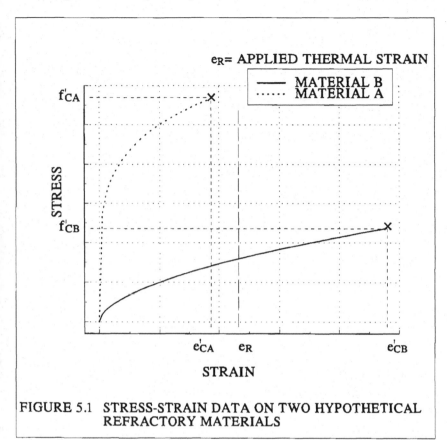

FIGURE 5.1 STRESS-STRAIN DATA ON TWO HYPOTHETICAL REFRACTORY MATERIALS

in addition to the gravity weight of the refractory lining itself.

In this example, the lining is also exposed to a uniform temperature of T_s. In this example, No restraint is placed on the thermal growth of the refractory. Therefore, no strain-controlled loading is imposed on the lining system. The operating stress for this example lining is simply defined as:

$$S_{sc} = P/A \qquad (5.1)$$

The stress is defined without the benefit of the mechanical material property data described in Figure 5.1. The full complement of the fundamental mechanical material property data would include the linear coefficient of thermal expansion (expressed as α), the modulus of elasticity (expressed as E), the bulk density (expressed as ρ) and Poisson's Ratio (expressed as v). The bulk density is used to define the stress-controlled gravity load. For the stress-controlled loading, only the structural geometry is required to evaluate the internal equilibrating stress within the structure. In the case of our simplified example, the cross-sectional area A is required. The internal equilibrating stress is evaluated for any structural system without the need of material properties. Even for highly redundant linear structures, only the structural geometry is required in defining the internal equilibrating stress when the structure is subjected to a stress-controlled load. To quantify our example, let us assume the load P is 1000 kg and the cross-sectional area A is 6.45 cc^2, the stress S_{SC}, substituting into Equation 5.1 would be:

$$S_{SC} = 1000/6.45 = 15.71 \text{ MPa (220 psi)}$$

With regard to the choice of strongest material, clearly refractory A is the obvious material to use for this application. Based solely on thermomechanical considerations, the choice is made by comparing the values of the ultimate compressive strengths of the two refractories. Refractory A has a much higher ultimate compressive strength than refractory B. If the degree of microfracturing is related to the ultimate compressive strength, then refractory A would be subjected to less microfracturing and the associated deleterious effects that shorten the life of a refractory lining. If long-term loading is applied, then creep effects would also be included in the choice.

In summary, for a stress-controlled loading, the compressive stress-strain curves for refractory A and B have no direct bearing on the evaluation of the choice of the strongest material. None of the mechanical material properties have a direct bearing on evaluating the stress in a structure subjected to a stress-controlled loading.

It should be pointed out that the primary emphasis is placed on the compressive stress-strain behavior of the refractory material. The reason being that most refractory lining systems are restrained to some degree against thermal growth and are, therefore, subjected to a compressive stress-strain environment. Tensile stress-strain environments are usually much less in magnitude due to two reasons. For linings consisting of brick shapes, no tensile forces can be transmitted across the brick joint, even when mortared joints are used. Secondly, refractory materials are much weaker in tension than in compression. The refractory lining, either a brick shape material or a castable material, cannot support a significant tensile stress-strain environment.

B. Choosing the Strongest Refractory for a Refractory System Subjected to Strain-Controlled Loading

In the second simplified example, the refractory dimensions are assumed to be identical to the first example. In the stress-controlled example, the refractory was heated to a temperature T_s. However, the lining was assumed totally unrestrained. Therefore, no expansion stresses were developed. In this second example (see Chapter 2, Figure 2.2), let us assume that an external support system restrains the full expansion of this lining system. In actual lining design, evaluating the magnitude of the restraint requires a thermal-structural analysis of the refractory lining system, including both the refractory lining and the external structural support system. This type of analysis is required in order to evaluate the thermal expansion interaction between the lining and the support. This analysis would require the mechanical material properties of both the support and refractory lining materials. In this simplified example, the refractory is assumed to be fully restrained.

Let us assume that both refractory A and refractory B had the same identical coefficient of thermal expansion. Therefore:

$$\alpha_A = \alpha_B \qquad (5.2)$$

Let us also assume, for the purposes of this example, that the steel support structure is very stiff and that both lining systems would be fully restrained. Therefore, the magnitude of the restrained thermal strain (e_R) on each of the two refractories is identical. This thermal strain is defined as:

$$e_R = \alpha(T_S - T_A) \qquad (5.3)$$

where T_A is ambient installed lining temperature.   Since both refractories have identical coefficients of expansion and are exposed to the same process temperatures, both refractory materials are subjected to identical thermal strains.   The thermal restraint strain in refractory A (e_{RA}) is identical to the restraint strain in refractory B (e_{RB}):

$$e_{RA} = e_{RB} \qquad (5.4)$$

It should be pointed out that the value of the thermal strains is achieved through a thermal-structural analysis that required both the geometry of the structure as well as the mechanical material properties.   Also, as defined elsewhere in this text, the thermal and structural analyses are done in two parts.   Part I is a thermal analysis, using thermal material properties to quantify the support and lining temperatures.   Part II is the structural analysis, using the Part I temperatures and the mechanical material properties to quantify the expansion interaction between the support structure and lining system.   The stress-strain behavior of both the support and lining are then obtained from the Part II analysis.

Returning to our example, the remaining quantity that has not been defined is the thermal restraint stress in each of the two refractory materials under investigation.   The thermal restraint stress (S_R) is calculated by multiplying the thermal strain by the modulus of elasticity:

$$S_R = e_R E \qquad (5.5)$$

The modulus of elasticity can only be defined through laboratory testing of a material.   In the case of steels, an ASTM tensile coupon is prepared and a tensile stress versus tensile strain curve is obtained, as illustrated in Figure 5.1.   It is the slope of this curve that defines the modulus of elasticity of the steel under investigation.   In the case of our two refractories, compressive test samples were prepared and tested at the defined operating temperature T_S.   The resulting compressive stress-strain curves for refractory A and refractory B are described in Figure 5.1.

Let us now evaluate the thermal restraint stress in each of the two materials.   Both materials are exposed to the identical thermal restraint strain. From Figure 5.1, projecting up from e_R to curve B, the thermal restraint strain results in a thermal restraint stress of S_{RB}, which is less than the ultimate compressive strength (f'_{CB}) of refractory B. However, for refractory A, the magnitude of restraint strain results in a thermal restraint stress (S_{RA}) that exceeds the ultimate compressive strength (f'_{CA}) of refractory A. In equation form, the two results are:

$$S_{RB} < f'_{CB} \qquad (5.6)$$

$$S_{RA} > f'_{CA} \qquad (5.7)$$

With an equal coefficient of expansion and equal process temperature, refractory A is inferior to refractory B. For these simplified example refractory lining systems, exposed to primarily a strain-controlled loading environment, and considering the thermomechanical aspects of the two materials, refractory B is superior to refractory A. Refractory A has superior ultimate compressive stress than refractory B but lower ultimate compressive strain (e'_C) than refractory B. In equation form:

$$e'_{CA} < e_R \qquad (5.8)$$

$$e'_{CB} > e_R \qquad (5.9)$$

Therefore, in the strain-controlled loading environment, refractory B is superior to refractory A.

In summary, the strain-controlled loading (restrained thermal expansion forces) requires the use of compressive stress-strain data in order to convert the thermal strain to a thermal stress. The compressive stress-strain data allows the refractory user to accomplish the primary design functions. First, it allows the refractory user to calculate the thermal stress for an imposed thermal strain condition. Without the compressive stress-strain data, the evaluation of the magnitude of thermal stress is impossible. Secondly, the compressive stress-strain data allows the refractory user to compare the calculated thermal stress with the ultimate crushing stress. The second design function is important in that it allows the user to make decisions with regard to the selection of the most appropriate refractory or with regard to the incorporation of expansion allowance to reduce the magnitude of the thermal stress-strain conditions. The compressive stress-strain data also allow the refractory user to compare the calculated thermal strain directly with the ultimate strain of the refractory materials. Typically, the structural analysis of the refractory lining system requires the compressive stress-strain data in which both the thermal strains and thermal stresses are a part of the structural analysis solution.

IV. VERIFICATION FROM FIELD TEST STUDIES

The scientific and technical reasoning for selecting the strongest refractory has been illustrated in the previous examples. There is presently no documented laboratory tests to verify this approach of selecting refractory material. The results of field tests are very limited [1]. The lack of both laboratory and field test experimental data can be attributed to the lack of understanding in the user and manufacturing communities on the impact of stress- or strain-controlled loadings on refractory lining structures. Or it can be attributed to the cost in developing the compressive stress-strain data.

An interesting field study was conducted by Dela Garza [2]. His work reflects the influence of ultimate crushing strain on the life of a refractory lining. Field test studies were conducted on the operating life of rotary kiln linings. The operating lives of several refractory manufacturers rotary kiln shapes were compared to two refractory kiln shapes.

V. SUMMARY

Since the majority of refractory lining systems serve as a heat containment system for a variety of industrial processes, it can be concluded that most refractory lining systems are subjected to strain-controlled loading and require a complete definition of the stress-strain behavior of the refractory material. With the lining subjected to thermal strains, the corresponding stresses can only be defined using the stress-strain data.

For the lining subjected to a stress-controlled loading, the strongest refractory should be selected on the basis of the greatest ultimate crushing strength and least amount of creep for long-term loading. For the lining subjected to a strain-controlled loading, the strongest refractory should be selected on the greatest ultimate crushing strain.

VI. RECOMMENDATIONS

The following recommendations [1] are made with regard to the structural and mechanical aspects of refractory use. These recommendations are intended to assist in optimizing these aspects of refractory lining systems. It should also be kept in mind that the chemistry aspects must also be addressed. The following recommendations are intended to be considered in addition to the chemistry aspects of refractory lining design.

The first recommendation is addressed to the refractory manufacturer. The second and third recommendations are addressed to the user of the refractory. The last recommendation is directed at both the user and manufacturing communities.

Recommendation No. 1

Since a majority of refractory lining systems are subjected to strain-controlled linings, it is desirable for the refractory manufacturer to provide to the user, the compressive stress-strain data [2] as part of the technical data on the supplied refractory material. The compressive stress-strain data are required by the user to assist in the decision-making process of selecting the appropriate refractory material. This data is an absolute necessity if the user's refractory lining system is exposed to a strain-controlled (thermal expansion) loading system.

Recommendation No. 2

The refractory user should, if the lining of interest is exposed to significant strain-controlled loadings, request compressive stress-strain data on the refractory materials from the refractory manufacturer. The ultimate crushing strength is adequate for refractory lining systems that are subjected to only a stress-controlled loading. However, for the lining system which is subjected to strain-controlled loading, the compressive stress-strain data are an absolute necessity for the user in selecting the best refractory for the application.

Recommendation No. 3

The refractory user that requires the compressive stress-strain data (has a lining system subjected to a strain-controlled loading) can use this data in two ways. First, the user can simply compare the compressive stress-strain data on the various refractories being considered and select the refractory material on the basis of the greatest strain range (the greatest ultimate crushing strain). Second, the user can conduct a structural analysis using the compressive stress-strain data for each refractory material under consideration. This second approach would determine if the refractory lining system is over-restrained, even for the most flexible material. The analysis would determine if the user's refractory lining system is in need of expansion allowance devices. The analytical approach would also identify if the user's lining system is exposed to a mild strain-controlled loading. Therefore, the stiffer refractory may be quite satisfactory.

Recommendation No. 4

Currently no ASTM specification exists that establishes the testing standards for laboratory development of the compressive stress-strain data. There are other similar materials, such as structural concrete, that have ASTM standards for conducting compressive strength tests. It is recommended that an ASTM standard be established by refractory users and manufacturing communities that sets standards for conducting compressive stress-strain tests. The standards would include:

a. The strain or stress rate for loading the sample. A strain rate would be more appropriate since we are addressing strain-controlled loading.

b. The sample size.

c. A verification on the compliance of the testing equipment based on well-known established materials.

d. A suggested coupling of the of the ultimate compressive stress and the corresponding ultimate compressive strain on the manufacturer's data sheets.

e. Based on the user's maximum process temperature imposed on the refractories, a suggested increment of testing temperatures, ranging from room temperature up to the user process temperature.

f. Minimum hold time at the specimen test temperature before the specimen is tested.

g. A procedure for preparing the sample (core drilling, etc.) and a method of preparing the ends to accept the compressive loading.

h. Core drilling direction based on users need and primary direction compressive loading.

i. Other requirements on core sample preparation such as core drilling speed [3] and other related conditions with regard to the sample preparation.

There will be, most likely, other requirements that would be imposed on the test procedure to assist in unifying the test results.

REFERENCES

1. Schacht, C. A., Recommended Additional Material Data for Evaluating the Mechanical Strength of Refractories, American Ceramic Society 89th Annual Meeting, April 26-30, 1987, Abstracts, p. 294.

2. Dela Garza, R., Importance of the Modulus of Elasticity on Basic Refractory Brick for Cement Rotary Kilns, American Ceramic Society, Annual Meeting May 5-9, 1985, ISSN 0196-6219, Vol. 7, No. 1-2, p. 284-292 (1986).

3. Faud, G. and Buyukozturk, O., Test Procedures and Interpretation of Results in Assessing Mechanical Properties of Refractories at High Temperature, American Ceramic Society, 90th Annual Meeting, 1988, Ceramic Transactions, Vol. 4, pp. 557-584 (1990).

6
Static Compressive Stress-Strain Data

I. INTRODUCTION

Perhaps one of the most important mechanical material property data for refractories is the compressive stress-strain data. Unlike the utimate compressive strength data, the compressive stress-strain data provide both the ultimate compressive strength and the matching ultimate compressive strain. With this material information, the refractory user can appropriately select the strongest refractory. In other words, based on the major type of lining loading (stress-controlled or strain-controlled) the user can choose the appropriate refractory for the lining system. More importantly, the compressive stress-strain data also provides a portion of the mechanical material properties required for structural analysis.

This chapter will define the compressive stress-strain data for a variety of refractory materials, including superduty fireclay brick, high-alumina brick, phosbonded high-alumina brick, fired mag-chrome brick, mag-carbon brick, a variety of castables, gunned mixes and other forms of refractories.

It is not the objective of this chapter to rate or rank the refractories presented with regard to strength or anyother aspects regarding the life of refractories. Rather, the presentation of the compressive stress-strain data is intended to provide insight for the refractory user in designing a refractory lining system. The availability of compressive stress-strain data is quite limited. In most cases, these data are not provided by the refractory manufacturer unless requested by the user. There are many instances in which users are unaware that such data are available or can be developed. The compressive stress-strain data are a fairly recent development compared to the other property parameters that are provided by the manufacturer. Therefore, the compressive stress-strain data are presented here, allowing the user to become familiar with this information. The compressive stress-strain data also provide considerable insight into refractory behavior at high temperatures.

Some refractories are inherently stiffer than others. Some appear to be extremely flexible. Obviously, the compressive stress-strain response of a refractory material will vary with chemistry, percent of impurities, basic material makeup, prefired manufacturing processing and firing details, the method of bonding and many other attributes of the manufacturing process. As a result, a user should obtain compressive stress-strain data from the manufacturer unique to the refractory under consideration.

II. BACKGROUND

The compressive stress-strain data described in this chapter originated from several laboratories equipped to develop these data. Each laboratory used test strain rates that were compatible with their own testing equipment. In addition, each laboratory differed in specimen size, specimen geometry, type of preparation equipment, type of refractory tested, method of specimen preparation, cutting or drilling rates in the specimen development and other details relating to the test environment.

For those familiar with the compressive stress-strain data, the compressive strain-strain data are quite often defined as *static compressive stress-strain* data. The time required to test a specimen lasts from several seconds to several minutes. Therefore, the time-dependent creep effects should be minimal. At higher temperatures (in excess of, say, 1400°C) creep does become significant, even for the short time periods of the test. It will be shown later, for the specimens tested at higher temperatures, that unaccounted creep effects may not greatly influence the accuracy of the solution to the refractory structural lining under consideration. The objectives of the refractory lining structure may differ between users. The reasoning for the inclusion or exclusion of creep effects should be based on these objectives.

The compressive stress-strain data are presented according to refractory types. In some cases the available data are very limited. In each refractory material presented, the manufacturer's defined chemistry is presented along with other associated data. In most cases, the ultimate crushing strength was established by the ASTM test procedures independent from the compressive stress-strain test. The data is presented in graphical form. In some instances, the polynomial equation that best fits the curve is also provided.

III. PRESENTATION OF COMPRESSIVE STRESS-STRAIN DATA

The following compressive stress-strain data on various types of refractories were developed by four different testing laboratories. In two instances the testing laboratory was also the refractory manufacturer. These two being Harbison-Walker Refractories Corporation (Pittsburgh, PA) and National Refractories and Minerals Corporation (Liverpool, CA). These two refractory suppliers also have the appropriate laboratory test equipment to develop compressive stress-strain data. The remaining two are independent laboratories: Babcock and Wilcox Corporation (Lynchburgh, VA) and Ceram Research (Stoke-on-Trent, Great Britain).

IV. FIRECLAY BRICK

Superduty fireclay compressive stress-strain data are described in Figure 6.1 for samples tested at 20, 260, 540, 815, 980 and 1150°C. This particular fireclay brick is no longer produced. However, the compressive stress-strain data provide valuable information on the mechanical behavior of quality superduty fireclay brick. As shown, when subjected to significant compressive stress, fireclay brick remains elastic for temperatures up to 815°C. At temperatures of 980 and 1150°C, plastic flow becomes significant. The compressive stress-strain data can also be expressed in polynomial form as:

$$S = C_1X + C_2X^2 + C_3X^3 + \cdots + C_nX^n \qquad (6.1)$$

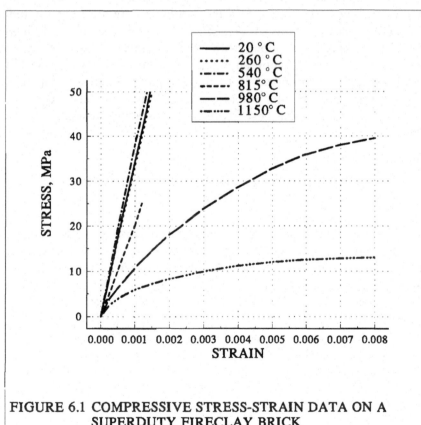

FIGURE 6.1 COMPRESSIVE STRESS-STRAIN DATA ON A
SUPERDUTY FIRECLAY BRICK

Table I summarizes the polynomial coefficients of the compressive stress-strain data in Figure 6.1.

Table I

Compressive Stress-Strain Data
of Figure 6.1*

TEMP. OF STRESS-STRAIN °C	POLYNOMIAL COEFFICIENTS, C_i x 10^n (EXPONENT n)						
	C_1	C_2	C_3	C_4	C_5	C_6	C_7
20	3.57 (4)**						
260	3.57 (4)**						
540	3.57 (4)**						
815	1.58 (4)	4.17 (6)					
980	1.49 (4)	-5.51 (6)	1.85 (9)	-3.5 (11)	3.27 (13)	-1.19 (15)	
1150	1.11 (4)	-8.20 (6)	4.04 (9)	-1.14 (12)	1.79 (14)	-1.47 (16)	4.86 (17)

*Blank table cell implies no coefficient.
**Straight line data for temperatures 20, 260, and 540°C nearly identical.

The manufacturer's [1] description of this superduty fireclay brick is summarized in Table II. The cold crushing strength is defined as ranging from 17.2 to 28 MPa in Table II. The cold crushing strength, as described in Figure 6.1, is in excess of 50 MPa. Additional manufacturer's data can be found in a more recent publication [2]. The compressive stress-strain curve for the 815°C temperature ends prematurely at about 25 MPa. This occurred most likely due to an abbreviated test or due to sample flaws developed during preparation. The strength at 815°C should range between the 540 and 980°C values.

Table II

Manufacterer's Description Of
Superduty Fireclay Brick [1]

CLASS	SUPERDUTY FIRECLAY BRICK
P.C.E. (Orton Cone)	33-34
Bulk Density, lb/ ft^3 (Kg/m^3)	144 to 148 (2300 to 2370)
Apparent Porosity, %	10.0 to 13.0
Cold Crushing Strength, lb/ in.^{2} (MPa)	2500 to 4000 (6.2 to 9.0)
Modulus of Rupture lb/ in.^{2} (MPa)	900 to 1300 (6.2 to 9.0)
Reheat Test, permanent linear change, %, after heating at 2910^{0}F (1600^{0}C)	-0.3 to -1.0
Load Test, 25 psi (172 kPa), % linear subsidence after heating at 2640^{0}F (1450^{0}C)	1.5 to 2.6
Panel Spalling Test, % loss, preheat at 3000^{0}F (1650^{0}C)	2.0 to 6.0

Chemical Analysis, approximate %
  Silica (SiO$_2$)
  Alumina (Al$_2$O$_3$)
  Titania (TiO$_2$)
  Iron Oxide (Fe$_2$O$_3$)
  Lime (CaO)
  Magnesia (MgO)
  Alkalies
  (Na$_2$O + K$_2$O + Li$_2$O)
	
	52.9
	42.0
	2.4
	1.0
	0.2
	0.3
	
	1.2

Note: These data are subject to reasonable variations and, therefore, should not be used for specification purposes.
ASTM test methods, where applicable, used for determination of data.
All data typical of brands at time of printing.

Table III

Compressive Stress-Strain Data
of Figure 6.2

TEMP. OF S-S CURVE, °C	POLYNOMIAL COEFFICIENTS, $c_i \times 10^n$ (EXPONENT n)									
	C_1	C_2	C_3	C_4	C_5	C_6	C_7	C_8	C_9	C_{10}
540	2.4 (3)	2.8 (6)	4.1 (10)	-5.0 (13)	2.6 (16)	-7.7 (18)	1.4 (21)	-1.4 (23)	8.2 (24)	-2.0 (26)
1090	3.3 (3)	1.9 (5)	-4.5 (7)	3.2 (9)	-1.1 (11)	1.9 (12)	-1.3 (13)			

Figure 6.2 describes the results of a separate compressive stress-strain test on the same superduty fireclay brick just defined. Here the compressive loading was continued until failure. The 540°C curve shows a maximum value of about 38 MPa. The linear portion of this curve reaches about 36 MPa at a strain of approximately 0.002. This is consistent with the 540°C curve in Figure 6.1.

The 1090°C compressive stress-strain curve of Figure 6.2 shows a continued increase in strength at strains of 0.04. A favorable comparison is observed with the 980°C curve of Figure 6.1. At a strain of 0.008, Figure 6.1 shows a strength of 40 MPa, while Figure 6.2 shows a strength of about 25 MPa. Considering the temperature differences, both test data appear realistic.

Each of the above test results were made from core-drilled samples of the fireclay brick. Considering the variability in the manufacturing process of refractory brick, some of variance between the two test results can be attributed to the sample differences.

V. HIGH-ALUMINA BRICK

High-alumina brick is a highly popular form of refractory used throughout industry. High-alumina brick belongs to the

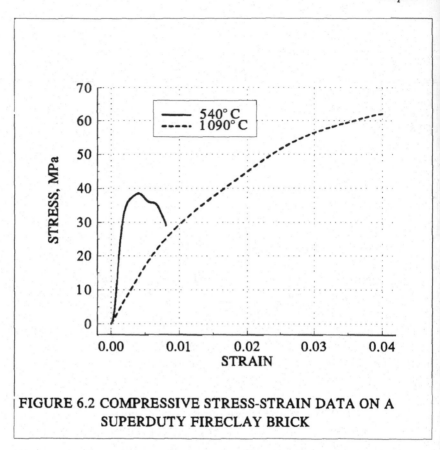

**FIGURE 6.2 COMPRESSIVE STRESS-STRAIN DATA ON A
SUPERDUTY FIRECLAY BRICK**

alumina-silica group [1,2] containing an alumina (A1203) content of
47.5% or higher.   The high-alumina title distinguishes them from
brick made of clay or other aluminosilicates in which the alumina is
below 47.5%.

The alumina classes of 50%, 60%, 70% and 80% have a
range of plus or minus 2.5%. The 85% and 90% classes have a
range of plus or minus 2%. The 99% class has only a minimum of
97% or only a minus range of 2%.

A. 50% Alumina Brick

The compressive stress-strain data of a fired 50% alumina

brick is described in Figure 6.3. For the purposes of this discussion, this 50% alumina brick is identified as Product No. 1 of three different 50% alumina brick products chosen for this discussion.

This 50% alumina brick is advertised to have a cold crushing strength of 52 MPa (7,540 psi). The alumina content is specified at 50.5%. The 538°C curve exhibits an apparent stiffening up to a load of about 30 to 40 MPa. Beyond the 40-MPa loading, the stress-strain response becomes fairly linear. The 1094°C curve is linear over the full tested range. Both samples exhibit considerable strength in excess of the advertised cold crushing strength.

Figure 6.4 describes the data [3] from a second 50% alumina

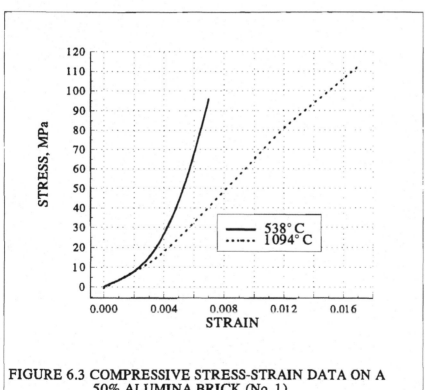

FIGURE 6.3 COMPRESSIVE STRESS-STRAIN DATA ON A
50% ALUMINA BRICK (No. 1)

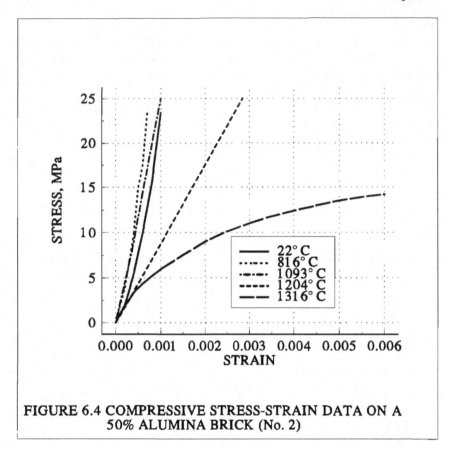

FIGURE 6.4 COMPRESSIVE STRESS-STRAIN DATA ON A
50% ALUMINA BRICK (No. 2)

brick product. This high-alumina brick is advertised to have a cold
crushing strength of 41 to 69 MPa. The compressive stress-strain
samples were only tested to a load level of about 25 MPa. The
alumina content is specified at 50.4%. Note that the lower
temperature curves (22 to 1093°C) show some stiffening, as
observed in Figure 6.3. The data in Figure 6.4 also show that
softening begins to occur at temperatures above 1093°C.
Significant softening occurs at temperatures above 1204°C.

        The third and final set of compressive stress-strain data of a
50% alumina brick is presented in Figure 6.5 [4]. According to the
manufacturer's data sheet, the advertised cold crushing strength is in
the range of 41 to 69 MPa. Type No. 3 is made by the same

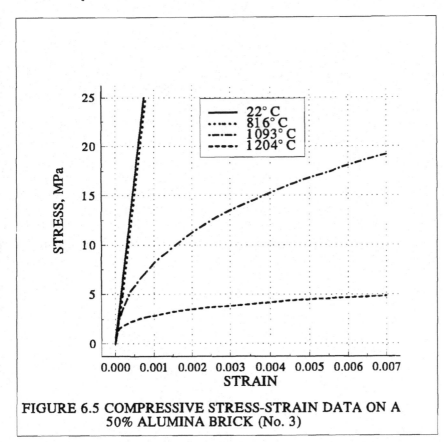

FIGURE 6.5 COMPRESSIVE STRESS-STRAIN DATA ON A
50% ALUMINA BRICK (No. 3)

manufacturer as Type No. 2. The alumina content is specified at
52.2%.

   In summary, at low temperatures the stress-strain data of
Product Nos. 2 and 3 reflect a somewhat stiffer refractory than
Product No. 1. At 1094°C, both Product Nos. 1 and 3 have similar
stiffnesses at strains up to about 0.004.   At high temperatures
(above 1093°C) Product No. 2 has considerably more stiffness than
Product No. 3, as observed by the 1204°C curve. Product No. 2 has
significant strength at 1316°C.

B. 70% Alumina Brick

The 70% alumina brick is the most frequently used high-alumina class through industry. Because of the popularity of this class of alumina brick, the compressive stress-strain data for five different 70% alumina brick products are presented. These data provide considerable insight into the structural behavior of 70% alumina brick and the brick parameters that influence the structural behavior.

The first of the 70% alumina brick products is presented in Figure 6.6 [5], though no specifics are provided with regard to the cold crushing strength, specific alumina content or other details. However, this 70% alumina brick appears typical for this class of alumina brick.

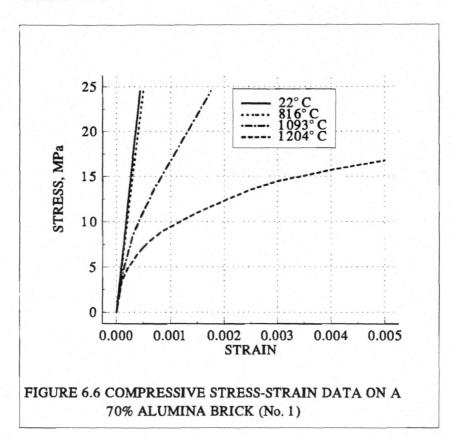

FIGURE 6.6 COMPRESSIVE STRESS-STRAIN DATA ON A
70% ALUMINA BRICK (No. 1)

Figure 6.7 describes another 70% alumina brick (No. 2) [4]. This brick has a lower specified density, in the range from 2643 to 2739 kg/m³ (165 to 171 lb/ft³), an alumina content of 72.1% and a cold crushing strength of 52 to 76 MPa (7,500 to 11,000 psi) [6].

The 70% alumina brick presented in Figure 6.8 (No. 3) [7,8] has the same brand name as the 70% alumina brick presented in Figure 6.7. Therefore the brick chemistry and other makeup parameters are also identical. The only difference between the types presented in Figures 6.7 and 6.8 is the density. The 70% alumina brick in Figure 6.8 was manufactured in the laboratory using a friction press. The result is a higher bulk density: 2755 kg/m³ (172 lb/ft³). The higher density is reflected in the greater stiffness observed in the compressive stress-strain data in Figure 6.8.

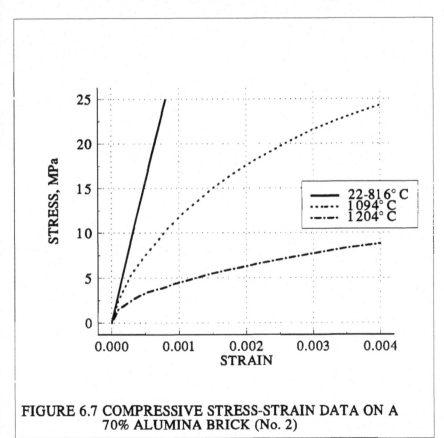

FIGURE 6.7 COMPRESSIVE STRESS-STRAIN DATA ON A
70% ALUMINA BRICK (No. 2)

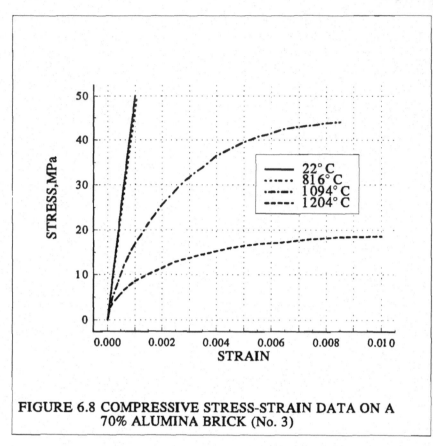

**FIGURE 6.8 COMPRESSIVE STRESS-STRAIN DATA ON A
70% ALUMINA BRICK (No. 3)**

The fourth 70% alumina brick product is described in Figure 6.9 [3,9,10]. This 70% high-alumina brick is defined to have a cold crushing strength of 28 to 41 MPa (4,000 to 6,000 psi). The alumina content is 72% and the bulk density ranges from 2500 to 2579 kg/m³ (156 to 161 lb/ft³). At temperatures of 1094 and 1204°C, the compressive stress-strain data of this product is quite similar to the stress-strain data of the alumina brick product of Figure 6.8. At low temperatures (22 to 816°C) the 70% alumina brick data in Figure 6.9 reflects a much greater flexibility than in the other three products. In Figure 6.9 the stress-strain curve at 22°C is curtailed at about 15 MPa. This is most likely attributable to a sample flaw or a malfunction of the testing equipment.

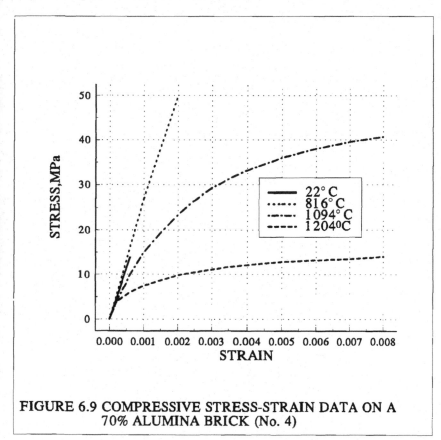

FIGURE 6.9 COMPRESSIVE STRESS-STRAIN DATA ON A
70% ALUMINA BRICK (No. 4)

Of particular interest is the stress-strain response of a chemically bonded 70% alumina brick (Product No. 5), described in Figure 6.10. The properties of this refractory brick are a bulk density equal to 2650 kg/m^3 (167 lb/ft^3), alumina content of 70% and a cold crushing strength of 66 MPa (9,500 psi). Compared to the stress-strain data of the previous fired 70% alumina bricks, this chemically bonded 70% alumina brick has considerably greater flexibility over the full range of temperatures considered.

The grouping of the compressive stress-strain data for this chemical bonded 70% alumina brick is also interesting. The 22 and 816°C data are very similar. Likewise, the 1094 and 1204°C are both similar. This implies that a significant change in stiffness

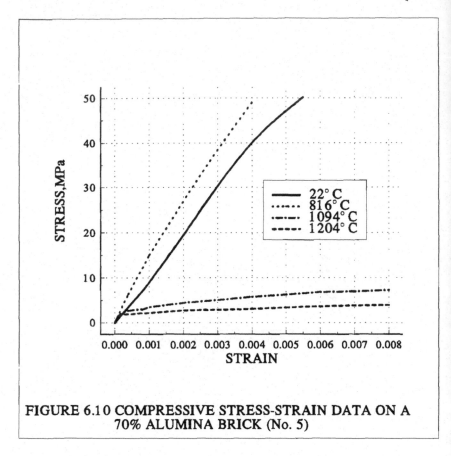

**FIGURE 6.10 COMPRESSIVE STRESS-STRAIN DATA ON A
70% ALUMINA BRICK (No. 5)**

occurs within the temperature range of 816 to 1094°C.

C. 80% Alumina Brick

The compressive stress-strain data on three different 80%
alumina brick products are examined. As expected, the greater the
alumina content, the greater the stiffness of the refractory brick.

Figure 6.11 is the compressive stress-strain data for the first
80% alumina brick Prduct (No. 1). This product is a phosphate-
bonded high-alumina brick. This product has an advertised cold
crushing strength of 143 MPa (20,750 psi) and an alumina content
of 82.6% [2]. Interestingly, the stress-strain data for the tempera-

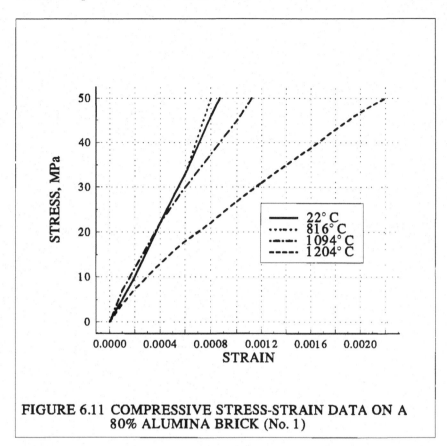

FIGURE 6.11 COMPRESSIVE STRESS-STRAIN DATA ON A
80% ALUMINA BRICK (No. 1)

tures of 22 and 816°C show more flexibility than the fired 70%
products. This can be attributed to the phosphate bonding. This
flexibility was also observed in the chemically bonded 70%
alumina brick. At the higher temperatures of 1094 and 1204°C, this
product exhibits considerable strength and flexibility.

The remaining two 80% alumina products (see Figures 6.12
and 6.13) are both fired alumina brick. The stress-strain [4] data
described in Figure 6.12 are for a domestic product, as were all of
the products described thus far. The stress-strain data described in
Figure 6.13 are for an 80% alumina brick product manufactured in
Australia [11].

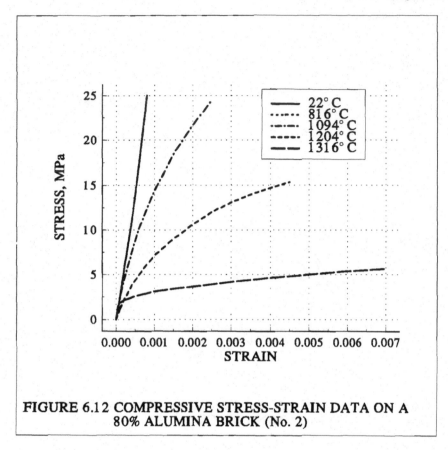

FIGURE 6.12 COMPRESSIVE STRESS-STRAIN DATA ON A
80% ALUMINA BRICK (No. 2)

It can be concluded that the 80% alumina brick has an
increased strength over the full temperature range of 22 to 1316°C.
Additional compressive stress-strain data on 80% alumina bricks
(85% based on Bauxite) with various types of bonding can be found
in Reference 12.

D. 90% Alumina Brick

The compressive stress-strain data on a 90% alumina brick
currently available are quite limited. Figure 6.14 describes
thecompressive stress-strain data on a fired 90% alumina brick
[6,10]. The specifications of the brick are 91.6% alumina and a cold
crushing strength ranging from 103 to 138 MPa (15,000 to 20,000

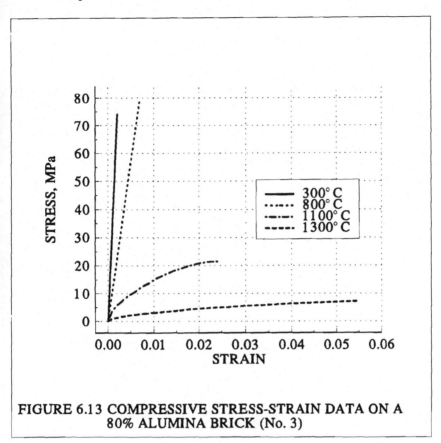

FIGURE 6.13 COMPRESSIVE STRESS-STRAIN DATA ON A
80% ALUMINA BRICK (No. 3)

psi). The ambient stress-strain data were only run up to a
compressive stress of just under 50 MPa, less than about half of the
ultimate crushing strength.

The influence of the 90% alumina content is shown by the
strength at the higher temperatures. Here significant stress-strain
response of the 90% alumina brick is exhibited at 1482°C.

E. Summary

The compressive stress-strain data have been presented for
a range of high-alumina brick products. The following trends are
observed with the range of high-alumina brick examined.

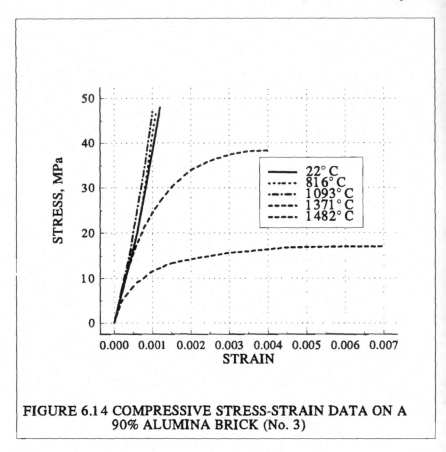

FIGURE 6.14 COMPRESSIVE STRESS-STRAIN DATA ON A
90% ALUMINA BRICK (No. 3)

The stiffness increases with increasing alumina content.
That is, for a given strain value, the stress increases with increasing
alumina content.

The stress-strain response becomes more significant as the
alumina content increases. The stiffness of the brick increases with
increasing alumina content.

In general, the chemically bonded high-alumina bricks have
more flexibility than the fired high-alumina bricks.

There appears to be other aspects of the high-alumina brick
chemistry that influence the compressive stress-strain response. It is

too premature to discuss these influences. Lime and alkali content appear to influence the stress strain response.

A portion of the high-alumina products experienced a strengthening or stiffening when samples were tested above 22°C. Testing at continuously higher temperatures showed that a peak strengthening or stiffening was reached, after which the samples became softer as temperatures increased. This phenomenon was seen, for example, in the 50% alumina brick Product No. 2. Although no investigations have been conducted to establish the cause, it is assumed that the increased temperatures result in either an increase in the bonding between various materials or an increase in the material strength.

Finally, the density appears to have a significant influence on the brick stiffness. As the density increases, the stiffness also increases.

Compressive stress-strain data on high alumina brick with an alumina content in excess of 90% were not available at the time of this writing. It can be expected, however, that as the alumina content increases to these higher percentages, the strength and stiffness at higher temperatures will also increase.

VI. MgO-CHROME BASIC BRICK

There is considerably less compressive stress-strain data available in the literature for basic brick than for high-alumina brick. However, the following basic brick compressive stress-strain data provide valuable information in comparing the stiffness of basic brick with high alumina brick. This type of information is greatly needed by the user to make judgments and decisions on the design of the basic refractory lining system.

Four basic brick products are used to illustrate the compressive stress-strain of MgO basic brick. Three of the basic products are 60% MgO dead burned basic brick. The fourth is a chemically bonded 60% MgO basic brick.

A. 60% MgO Dead Burned Brick Product No. 1

Figure 6.15 describes the compressive stress-strain data on a 60% MgO dead burned (DB) basic brick [13]. The reference does not cite the product name nor the cold crushing strength. Literature on other similar basic brick suggests that the cold crushing strength may range from 28 to 59 MPa (4,000 to 8,500 psi). An interesting characteristic of the 60% MgO DB brick, when compared to the previously discussed high alumina brick, is the stiffness of the subject basic brick for temperatures up to 1300°C. The high-alumina brick (up to 90% alumina) exhibited loss of stiffness at temperatures above about 1300 to 1400°C. The basic brick, as described in Figure 6.15, exhibits stiffness at temperatures of 1300°C equal to the stiffness at room temperature.

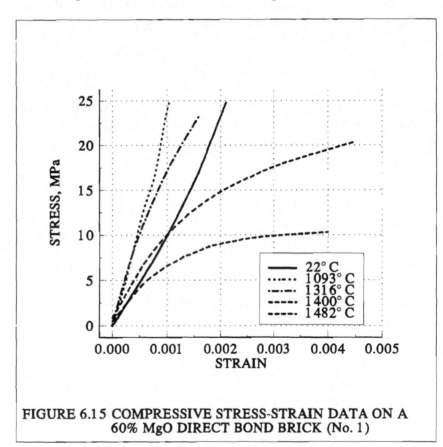

FIGURE 6.15 COMPRESSIVE STRESS-STRAIN DATA ON A
60% MgO DIRECT BOND BRICK (No. 1)

B. 60% MgO DB Brick Product No. 2

The second set of 60% MgO DB basic brick compressive stress-strain data are described in Figure 6.16 [5]. Just as with No. 1, no cold crushing strength was cited for this basic brick. This basic brick also exhibits a similar stiffness over the full temperature range of 22 to 1316°C. At 1316°C, some departure is shown at the higher levels of compressive stress, where it exhibits a greater flexibility. It would be expected that the cold crushing strength of this basic brick should be in the range of 20 to 59 MPa.

C. 60% MgO DB Basic Brick Product No. 3

The third set of 60% MgO DB basic brick product stress-

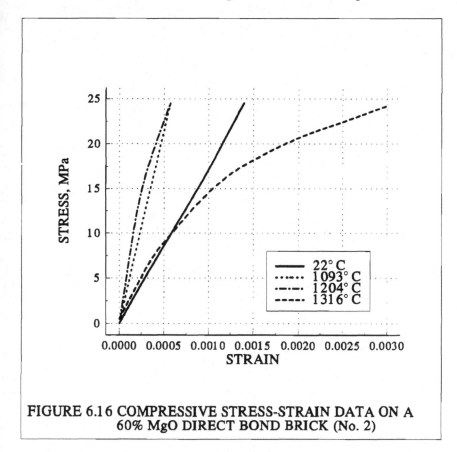

FIGURE 6.16 COMPRESSIVE STRESS-STRAIN DATA ON A
60% MgO DIRECT BOND BRICK (No. 2)

strain data (see Figure 6.17) [4,10], show considerable flexibility at temperatures in excess of 1316°C. This basic brick is 62.4% MgO and has a cold crushing strength of 69 MPa (10,000 psi). The room temperature stress-strain data were only developed up to a compressive stress of 25 MPa. Similar stiffnesses are seen at the lower temperature regimes, as in the previous two basic products. Significant high-temperature stiffness of the product is illustrated by the 1482°C stress-strain data.

D. 60% MgO Chemically Bonded Basic Brick Product No. 4

The fourth 60% MgO basic brick product is chemically bonded (see Figure 6.18) [4]. At temperatures of 1093°C and higher, the subject chemically bonded basic brick has considerably more

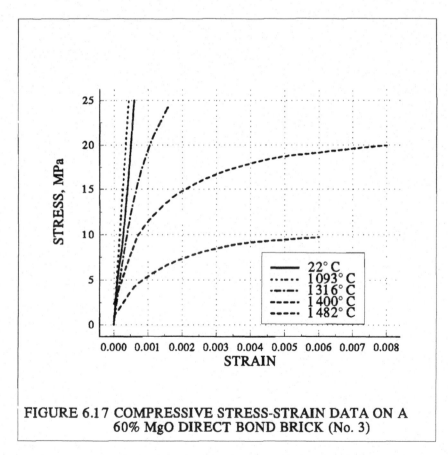

FIGURE 6.17 COMPRESSIVE STRESS-STRAIN DATA ON A 60% MgO DIRECT BOND BRICK (No. 3)

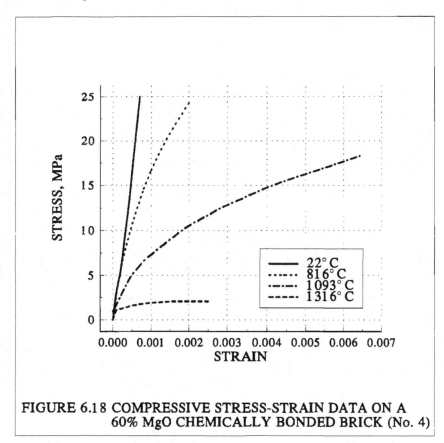

FIGURE 6.18 COMPRESSIVE STRESS-STRAIN DATA ON A
60% MgO CHEMICALLY BONDED BRICK (No. 4)

flexibility than the DB basic bricks. The MgO content and cold
cold crushing strength are not identified. The MgO content is
expected to be near that of the previous basic products. The cold
crushing strength is not expected to be above the range of the
previous basic products.

E. Summary

In general, for the same temperature range, the stiffness of
the 60% MgO DB basic brick is greater than the 70% alumina
brick, for alumina contents approaching 90%. The 60% MgO DB
basic brick has similar stiffness at temperatures above 1400°C. This
information has significant impact on the design of alumina and

basic lining systems.

VII. MgO-CARBON BASIC BRICK

Stress-strain data of MgO-carbon brick is very limited due to the complexity in conducting the compressive stress-strain tests at high temperatures, especially above 1000°C where inert nonoxidizing environments are difficult to maintain. The compressive stress-strain data [14] for two MgO-carbon products are presented.

A. MgO-Carbon Resin Bond Brick

Figure 6.19 describes the compressive stress-strain data of a MgO-carbon resin bond brick. This brick has a 20% carbon content with metal (alumina) additions. The upper temperature limit of the stress-strain data is 1093°C (2000°F). Both the 22°C and 1093°C curves have a similar response. Therefore, this product is not significantly affected by temperature, at least within the 22 to 1093°C range. It is expected, however, that this product would show softening for the stress-strain data above 1093°C. Compared to the 60% MgO direct bond and chemically bonded products, this MgO-carbon brick is fairly soft. As the compressive stress increases, the inelastic flow also increases.

Comparison of the MgO-carbon resin bond brick with the previous MgO-chrome direct bond brick (see Figures 6.15 to 6.17) shows that the MgO-carbon resin bond brick is more flexible, especially at room temperature.

B. MgO-Carbon Pitch Bond Brick

Figure 6.20 describes the compressive stress-strain data of a MgO-carbon pitch bond brick. This brick has a 5% carbon content and does not have any metal additions.

Comparison of the two MgO-carbon bricks (Figures 6.19 and 6.20) indicates that both of these products have similar ranges of stiffness at the higher temperatures. The 5% carbon brick does exhibit a greater stiffness at room temperature. Since thermal expansion reaction forces in vessel linings are not present at room temperature, the room temperature stress-strain data are of little

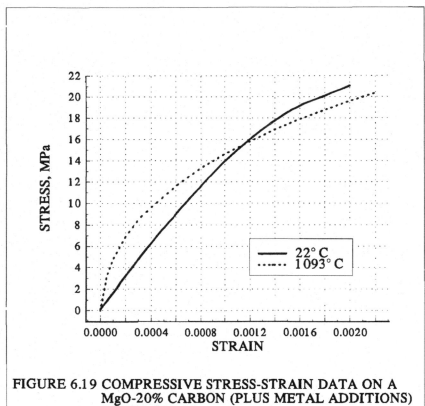

FIGURE 6.19 COMPRESSIVE STRESS-STRAIN DATA ON A
MgO-20% CARBON (PLUS METAL ADDITIONS)
BRICK (NO. 1)

consequence in the lining design, except that it does reflect the degree of softening as a function of temperature.

C. Summary

Even with state-of-the-art testing equipment, inert environments cannot be developed for investigating the compressive stress-strain data on MgO-carbon refractory products. For this reason the compressive stress-strain data on MgO-carbon products presented here are limited to 1093°C.

The stiffnesses of both MgO-carbon products (No. 1: resin bond with metal additions and No. 2: pitch bond) are similar

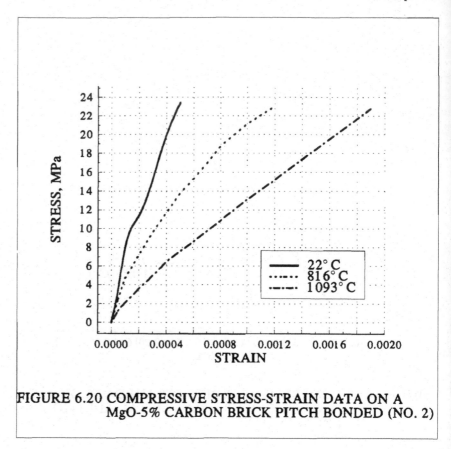

FIGURE 6.20 COMPRESSIVE STRESS-STRAIN DATA ON A
MgO-5% CARBON BRICK PITCH BONDED (NO. 2)

at temperatures up to about 1093°C.

It is expected that both of the mag-carbon products will
continue to soften at higher tempreature since MgO will soften, as
evidenced in the 60% MgO DB products.

VIII.   DOLOMITIC BRICK

Dolomitic brick has continued to gain popularity in many
industrial applications. The following stress-strain data are expected
to provide the necessary information for designing a dolomitic
lining and to optimize the thermomechanical behavior of the
dolomitic lining system. The following stress-strain data [15] are

for two popular forms of dolomitic brick, direct bonded and resin bonded.

A. Dolomitic Direct Bonded Brick

The mechanical properties of this form of dolomitic brick are provided in the manufacturers data sheet [16]. The coefficient of thermal expansion for this type of refractory brick is 4.1 x 10^{-6} mm/mm°C (7.4 x 10^{-6} in./in.°F). The compressive stress-strain data for this dolomitic brick (No. 1) are shown in Figure 6.21. The strain, for the temperature range of 1093 to 1316°C, is quite significant for a direct bonded brick.

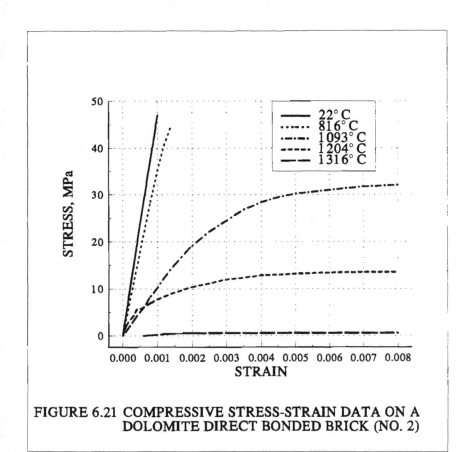

FIGURE 6.21 COMPRESSIVE STRESS-STRAIN DATA ON A
DOLOMITE DIRECT BONDED BRICK (NO. 2)

B. Dolomitic Resin Bonded Brick

The mechanical property of this form of dolomitic brick (No. 2) is provided in the manufacturer's data sheet [17]. The coefficient of thermal expansion for this type of dolomite brick is 3.56×10^{-6} mm/mm°C (6.4×10^{-6} in./in.°F). The compressive stress-strain data for this dolomite brick are described in Figure 6.22. Because of the test limitations, the compressive stress-strain data could only be evaluated for up to 1093°C. With respect to stiffness, the resin brick has a more gradual transition from room temperature up to 1093°C. There is a reversal of stiffness, however, as shown by the 536 and 816°C data curves. The resin bonded brick becomes less stiff at 538°C than at 816°C. This may be due to the temperature-dependent behavior of the resin.

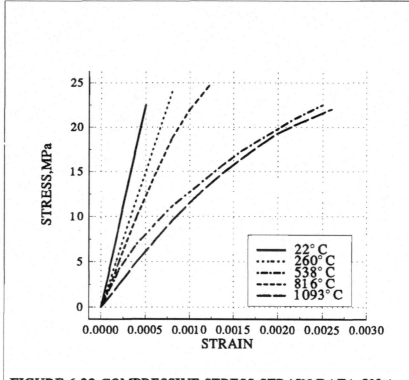

FIGURE 6.22 COMPRESSIVE STRESS-STRAIN DATA ON A
DOLOMITE RESIN BONDED BRICK (NO. 2)

C. Summary

As with the MgO-carbon brick, the current state-of-the-art testing equipment limits the stress-strain data of resin bonded brick to about 1093°C. Most likely, the resin bonded dolomitic brick will continue to soften at temperatures above 1093°C.

IX. SILICA BRICK

Figures 6.23 and 6.24 describe the compressive stress-strain data on two different silica brick (18,19). No other mechanical property data were available for these two silica refractory brick materials. The silica brick stress-strain data in Figure 6.23 exhibit considerably more strain range for the 1200°C data curve than at

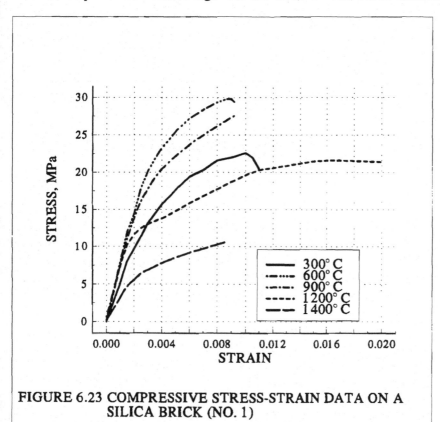

FIGURE 6.23 COMPRESSIVE STRESS-STRAIN DATA ON A
SILICA BRICK (NO. 1)

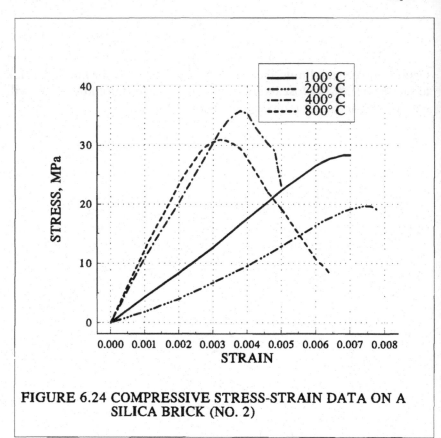

FIGURE 6.24 COMPRESSIVE STRESS-STRAIN DATA ON A
SILICA BRICK (NO. 2)

the other temperatures. The silica brick data described in Figure
6.24 were developed for temperatures up to 800°C. Neglecting the
1200°C data curve, both silica brick exhibit similar strain ranges.
The silica brick of Figure 6.24, however, exhibits a loss of strength
at 400 and 800°C for the strains in excess of about 0.003 to 0.004.

One interesting phenomenon observed in both silica brick
data is the variation in stiffness above room temperature. The data
in Figure 6.23 reflects an increase in stiffness above the 300°C
temperature and a gradual reduction for temperature above 600 to
900°C. In Figure 6.24, an increase in stiffness is observed
for temperatures above 200°C. Because of the test limitations on
temperature the decrease in stiffness at the high temperature could

not be identified. Naruse and Hoshind [20] reported a similar trend in silica brick defined as dense and superdense quality silica brick. Taking the slopes of the stress-strain data (slope = modulus of elasticity, MOE) they plotted the MOE as a function of temperature as shown in Figure 6.25. Their data is in close agreement with the independent stress-strain data shown in Figures 6.23 and 6.24.

The preceding data imply that there is an intermediate temperature range in which the silica brick tends to soften.

X. CASTABLES

There is considerably less compressive stress-strain data on

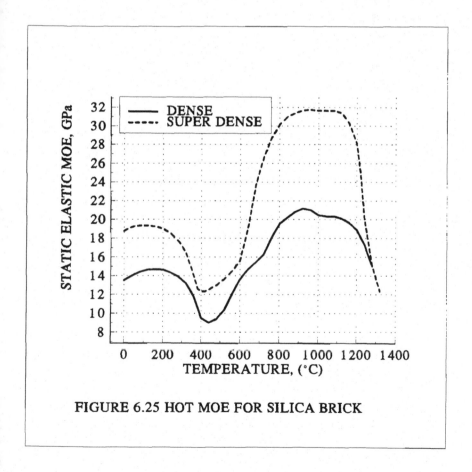

FIGURE 6.25 HOT MOE FOR SILICA BRICK

refractory castables than for the various refractory brick. The following data should provide considerable insight into mechanical behavior of castables and into the behavior of monolithic lining systems.

A. Alumina Castable

The following compressive stress-strain data are for a castable with an alumina content of 57.94%, according to the manufacturer's specifications [6]. This castable also has a crushing strength ranging from 26 to about 55 MPa over the temperature range of 100 to 1315°C. As in most castable refractories, there is shrinkage that takes place between the *as cast* (wet in mold or *green*) state in drying and final heated condition. In some cases, the shrinkage can be significant and will offset the thermal expansion of the castable. With regard to mechanical properties, the shrinkage of castables is perhaps one of the distinguishing differences from those of fired refractory brick.

A most interesting factor that has been included in the following castable stress-strain data [21] is the influence of the installation procedure on the stress-strain data. The stress-strain data in Figure 6.26 are for the vibration cast form of the subject castable. Note the increased stiffness for the temperature range of 22 to 1093°C when compared to the gunned installation (see Figure 6.27). Basically, the vibration cast-installed castable would be significantly more dense, or less porous. The density can be defined as one of the parameters used to rank the stiffness of a given castable.

The incorporation of steel fibers does not result in an effective stiffening of the castable. The additional compressive stress-strain data [21] on the same above castable (vibration cast) with steel fibers did not differ significantly from those shown in Figure 6.26. Close comparison shows that the steel fiber addition causes a slight softening of the castable at temperatures above room temperature. This is attributed to the difference in thermal expansion between the steel fibers and the surrounding castable. The fibers can function properly if there is a shear bond at the interface of the fiber and castable, as in reinforced structural concrete.

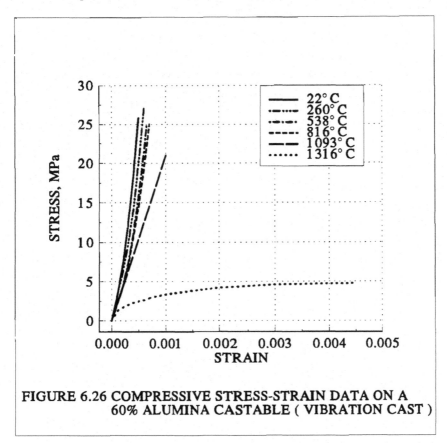

FIGURE 6.26 COMPRESSIVE STRESS-STRAIN DATA ON A
60% ALUMINA CASTABLE (VIBRATION CAST)

The heatup temperature results in the fiber expanding more than
the castable, causing deterioration of this shear bond. Most likely
the castable at the two ends of fiber is also crushed. The result is
that local deterioration of the castable may occur in the vicinity of
the fiber ends and the castable surrounding the fiber.

B. 45% Alumina Castable

The following compressive stress-strain data reflect a
chemical influence on the stress-strain behavior of an alumina
castable. Figures 6.28 and 6.29 describe compressive stress-strain
data for a conventional 45% alumina castable and a 45% alumina
low-cement castable, respectively [5]. Even with a low-alumina

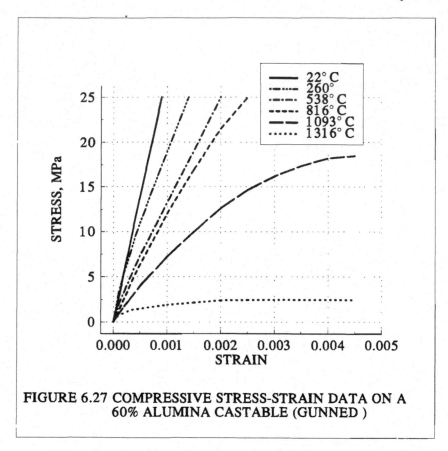

FIGURE 6.27 COMPRESSIVE STRESS-STRAIN DATA ON A
60% ALUMINA CASTABLE (GUNNED)

content, alumina castable can exhibit considerable stiffness with a
low cement content.   Exclusive of cement content, the percent
alumina is expected to influence the   stiffness of the alumina
castable. In general, the higher the percent alumina, the stiffer the
alumina castable.   Comparison of the low-cement 45% alumina
castable data (Figure 6.29) with the gunned 60% alumina castable
(Figure 6.27) reflects the influence of the cement content.

C. Low Cement 70% Alumina Castable

The compressive stress-strain data on a low-cement 70%
alumina castable (normal casting procedure) are described in Figure
6.30 [5]. According to the manufacturer's data sheet [22], this

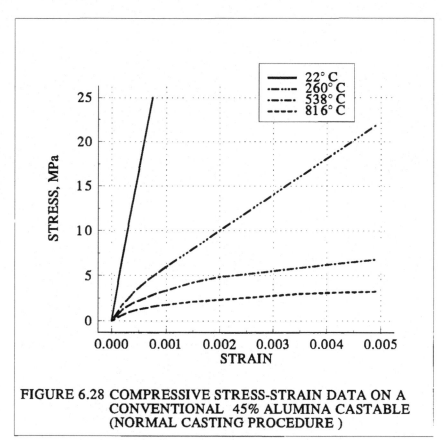

FIGURE 6.28 COMPRESSIVE STRESS-STRAIN DATA ON A
CONVENTIONAL 45% ALUMINA CASTABLE
(NORMAL CASTING PROCEDURE)

castable has a cold crushing strength ranging from 41 to 47 MPa at
room temperature. Comparing this low-cement 70% castable with
the low-cement 45% castable (see Figure 6.29) shows the higher
alumina content can have a significant influence on the castable
stress-strain behavior. The percentage of cement content for these
low-cement castables has not been determined here.

D. 94% Alumina Castable

The compressive stress-strain behavior of a 94% alumina
castable is shown in Figure 6.31. The comparison of the stiffness
of this conventional 94% alumina castable with a conventional 45%
alumina (see Figure 6.28) shows the impact of the alumina content.

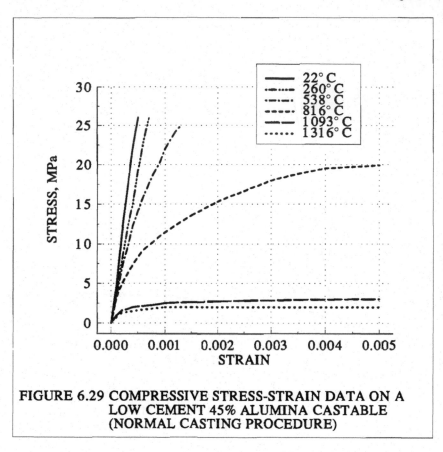

FIGURE 6.29 COMPRESSIVE STRESS-STRAIN DATA ON A
LOW CEMENT 45% ALUMINA CASTABLE
(NORMAL CASTING PROCEDURE)

According to the manufacturer's data sheet [2], this castable has a
cold crushing strength (after drying at 110°C) of about 58 MPa. It
should also be noted that the 94% castable has a significant strain
range for temperatures above 1093°C.

E. 98% Alumina Castable

Figure 6.32 describes the compressive stress-strain data on a
98% alumina castable [22]. The manufacturer's data sheet
describes a cold crushing strength for this castable that ranges from
35 to 41 MPa. Note that significant stiffness and strain range is
exhibited at temperatures above 1370°C.

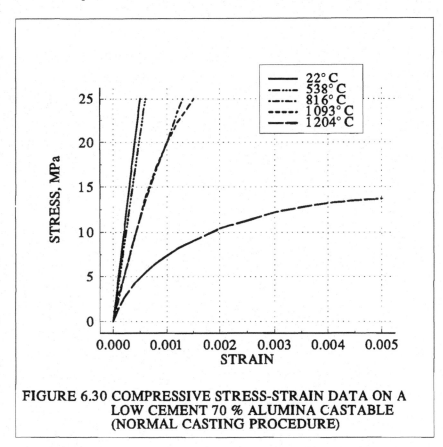

FIGURE 6.30 COMPRESSIVE STRESS-STRAIN DATA ON A
LOW CEMENT 70 % ALUMINA CASTABLE
(NORMAL CASTING PROCEDURE)

F. Summary

Unlike the refractory brick, the refractory castable is basically manufactured (installed) at the sight of the process vessel. In some instances castable shapes can be installed at the manufacturer's plant sight. The density of the refractory brick is controlled by the brick press. The brick press is typically a mechanical press. In some instances they are hydraulic or friction types. Therefore, the brick is made at a controlled density [23]. The castable, on the other hand, can have a variation in density based on the installation method.

The chemistry of castables has a significant influence on the

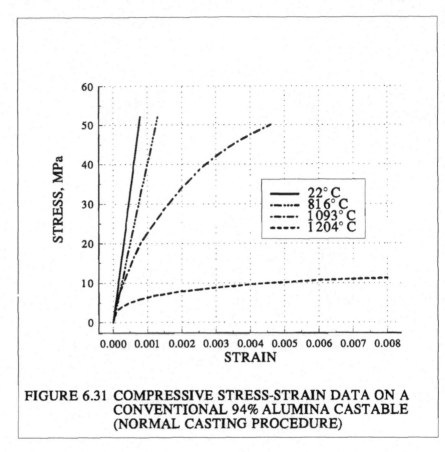

FIGURE 6.31 COMPRESSIVE STRESS-STRAIN DATA ON A
CONVENTIONAL 94% ALUMINA CASTABLE
(NORMAL CASTING PROCEDURE)

stress-strain behavior. Typically, the lower the cement content of
alumina castables, the greater the stiffness; and the higher the
alumina content, the greater the stiffness.

The addition of steel fibers to a castable does not have
a significant influence on the stiffness of the castables. The steel
fibers (either carbon or stainless steel) expand to a greater degree
than the castable, causing a highly localized deterioration of the
castable. As a result most test data show that the use of steel fibers
tends to reduce the strength at high operating temperatures [24],
while increasing the strength at lower temperatures.

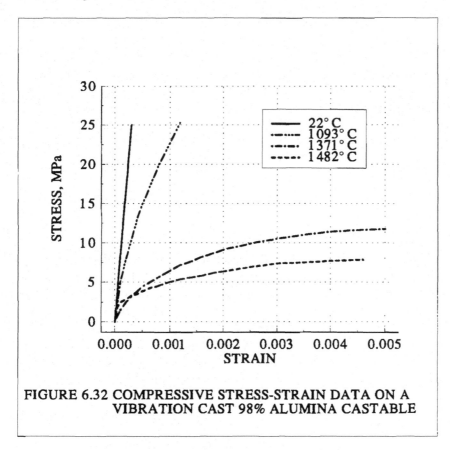

FIGURE 6.32 COMPRESSIVE STRESS-STRAIN DATA ON A
VIBRATION CAST 98% ALUMINA CASTABLE

XI. MATHEMATICAL DEFINITIONS OF THE
STRESS-STRAIN RELATIONSHIP

As previously discussed in Chapter 3, considerable work has
been conducted on structural concrete in defining the stress-strain
behavior, strength behavior and other relationships related to how
concrete materials behave. The following discussion addresses the
various mathematical forms used to quantify the stress-strain
relationships for structural concrete. The limitation on this
information is the lack of temperature dependency.

The use of an equation form of the stress-strain data is most applicable for computer analysis of refractory lining systems. The stress-strain data from laboratory tests are also used for computer analysis. Some structural computer programs require a definition of parameters taken from the stress-strain data sheets such as the initial tangent modulus. Some computer programs require the stress-strain data in equation form.

One form of mathematical definition [25] of the stress-strain relation for structural concrete is of the dimensionless form:

$$y = mx / \{ 1 + [m - (n / (n - 1))] x + x^n / (n - 1) \} \qquad (6.2)$$

where $y = f_c / f'_c$, the ratio of the concrete compressive stress to the ultimate strength; $x = \epsilon_c / \epsilon_o$, the ratio of the concrete strain to the strain at $y = 1$ (the ultimate strain) ; $m = E_o / E_s$, the ratio of the initial tangent modulus to the secant modulus at $y = 1$ (at ultimate stress); and n is a factor to control the steepness rate for the descending portions of the stress-strain relation. As the strain increases in the descending portion of the stress-strain curve, m controls the slope of the descent. Figure 6.33 describes the stress-strain relation for a concrete material using Equation 6.2.

Concrete stress-strain has been expressed in the following form as [26]:

$$f_c = (A\epsilon_c + B\epsilon_c^2) / (1 + C\epsilon_c + D\epsilon_c^2) \qquad (6.3)$$

where the stress-strain data are expressed as the ratio of the two quadradics. The constants A, B, C and D are determined from a curve fit of the test data. Using Popovic's [27] suggested definitions, the parameters m and n for equation 6.2 are defined as [25]:

$$m = 1 + 17.9 / f'_c \qquad (6.4)$$

$$n = f'_c / 6.68 - 1.85 > 1 \qquad (6.5)$$

Substituting into Equation 6.2, the resulting stress-strain curve, as described in Figure 6.33, is obtained for concretes with a range

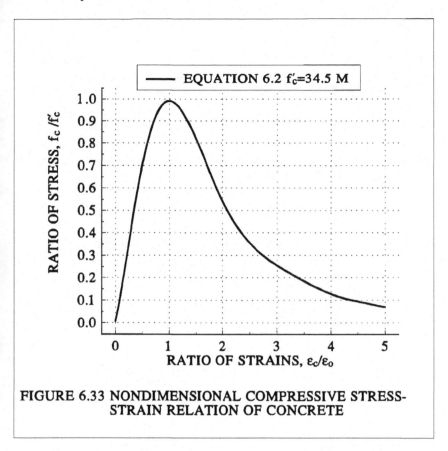

FIGURE 6.33 NONDIMENSIONAL COMPRESSIVE STRESS-
STRAIN RELATION OF CONCRETE

of f'_c. The Equation 6.2 formulation agrees well with forms defined by Wang and Popovics. Equations 6.4 and 6.5 are in MPa units.

Another form of compressive stress-strain relationship for structural concrete is described as [28]:

$$f_c = \epsilon_c E_o / \{ 1 + [(E_o / E_s) - 2][\epsilon_c / \epsilon'_c] + [\epsilon_c / \epsilon'_c]^2 \} \qquad (6.6)$$

where the ultimate strain (ϵ'_c) is:

$$\epsilon = (31.5 - \{f'_c\}^{0.25})(f'_c \times 10^{-5})^{0.25} \qquad (6.7)$$

Equation 6.6 (see Figure 6.34) is not applicable for values of E_s/E_o greater than 0.5 (or E_o/E_s less than 2). That is, if E_s is of suffient magnitude that it is greater than half the value of E_o, the curve is not considered to be significantly nonlinear and Equation 6.6 is not applicable. Equation 6.7 is in psi units.

These formulations were modified for application to castable refractory by accounting for the influence of temperature [29]. For the linear stress-strain behavior (typical at lower temperature and lower strains), the following equation was used:

$$\text{If } \epsilon_c \, / \, \epsilon'_c \; < \; 2 - E_o/E_s \; \text{ then use } \; f_c = E_o \, \epsilon_c$$

Beyond the linear stress-strain range (usually at higher strains and

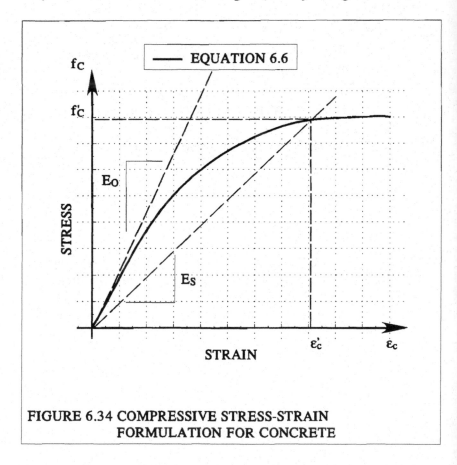

FIGURE 6.34 COMPRESSIVE STRESS-STRAIN
FORMULATION FOR CONCRETE

higher temperatures) if $\epsilon_c / \epsilon'_c > 2 - E_o / E_s$ then use Equation 6.6.

For the castables, the initial tangent modulus (E_o) decreased with increasing temperature. To account for the temperature effects, the tangent modulus was modified as:

$$E_o = E_{oR} \{ 1 + A [e^{-B(T + 70)} - 1] \} \qquad (6.8)$$

where E_{oR} is the initial tangent modulus at room temperature. The constants A and B are obtained by a curve fit of the data.

REFERENCES

1. Harbison-Walker Handbook of Refractory Practices, Second Edition, Harbison-Walker Refractories, Division of Dresser Industries, Inc. (One Gateway Center, Pittsburgh, PA 15222) p. 25, 1980.

2. Harbison-Walker Handbook of Refractory Practice, Harbison-Walker Refractories, Division of INDRESCO, Inc. (One Gateway Center, Pittsburgh, PA 15222) PD-9, TD-19, 1992.

3. Refractories for the Aluminum Industry, Material Specifications, National Refractories & Minerals Corporation, Livermore, CA, circa 1983.

4. Hartsock, M. A., Stress-Strain Data on Ladle Brick, Technical Memorandum, Kaiser Refractories Center for Technology, Kaiser Aluminum & Chemical Corp., July 6, 1983.

5. Alder, W. R. and Masaryk, J. S., Elevated Temperature Stress-Strain Measurements on Refractories, 2nd International Conference on Refractories, Tokyo, Japan, Nov., 1987.

6. Refractories Research, Technical Data, Kaiser Aluminum and Chemical Corporation (National Refractories & Minerals

Corporation), Livermore, CA, 1981.

7. Schacht, C. A., AISE Ladle Study SC-9-1, AISE Interim Reports No. 1 through No. 4 and Final Report, September, 1984.

8. Hartsock, M. A., MTS Test of Krial 70 HS (172 PCF), Laboratory Services Report, Kaiser Aluminum & Chemical Corp., Feb., 1984.

9. Alder, W. R. Kricor and Alumex 70-E Stress-Strain Data, Technical Memorandum, Kaiser Aluminum & Chemical Corp.(National Refractories & Minerals Corporation) Livermore, CA , March 30, 1983.

10. Alder, W. R. and Masaryk, J. S., Compressive Stress-Strain Measurement of Monolithic Refractories at Elevated Temperatures, Amer. Cer. Soc., Proceedings, 13, pp. 97-109 (1985).

11. Hialex 805 Data Sheet Australian Industries Refractories, BHP Steel, Newcastle, Australia, March 1990.

12. Padgett, G. C., Cox, J. A. and Clements J. F., Stress/Strain Behavior Materials at High Temperatures, Trans. Brit. Ceram. Soc. 68 (2), pp. 63-72 (1967).

13. Stett, M. A., Measurement of Properties for Use with Finite Element Analysis Modeling, Amer. Cer. Soc., Proceedings, Ceramic Eng. and Science, Applications of Refractories, ISSN 0196-6219, pp. 196-208 (1986).

14. Farris, R. E. and Schacht, C. A., Properties of MgO-Carbon Brick and Their Effect on Lining and Shell Stresses in a BOF, ISS Fiftieth Ironmaking Conference and Seventy-Fourth Steelmaking Conference, 1991.

15. Schacht, C. A., Investigative Study on the Thermomechanical Behavior of Resin-Bonded Dolomitic Brick Working Lining, AISE Mini-Study, March, 1987.

16. Baker Dolomite Refractory Products, Specification Sheet, York Fired-L (18), bg/07611-SM/49, Std., August, 1987.

17. Baker Dolomite Refractory Products Specification Sheet, Dolomax RBD, bg/07611-SM/5, Std., August, 1987.

18. Nineteenth Report on the Work of the Refractories Division 1978/79, British Cer. Res. Assoc., Oct., 1979, Penkhull, Stoke-on-Trent, England.

19. Stress Calculations for the Sections of No. 2A Blast Furnace Stove at Altos, Hornos De Vizcaya, S. A. Constructed with Silia Refractories, British Cer. Res. Assoc., Sept., 1984, Penkhull, Stoke-on-Trent, England.

20. Naruse, Y. and Hoshino, Y., Silica Bricks for Coke Ovens, Taikabutsu Overseas, Vol. 2, No. 1, pp. 110-120(1982).

21. Alder, W. R., Stress/Strain Testing of Refractories for the Hydrocarbon Processing Industry, Kaiser Refractories (National Refractories & Minerals Corporation) (1983).

22. Material Specifications for Kricon 32-70 and Kricon 34, National Refractories and Minerals Corporation, Livermoore, CA.

23. Schacht, C. A., A Viscoelastic Analysis of a Refractory Structure, Ph.D. Thesis, Carnegie-Mellon University April, 1972.

24. Lankard, D. R. and Sheets, H. D., Use of Steel Wire Fibers In Refractory Castables, Amer. Cer. Soc., Ceramic Bulletin, Vol. 50, No.5, pp. 496-500, 1971.

25. Tsai, W. T., Uniaxial Compressive Stress-Strain Relation of Concrete, J. Struct. Eng., ASCE, 114(9) pp. 2133-2136 (1988).

26. Wang, P., Shah, S. and Naaman, A., High-Strength Concrete in Ultimate Strength Design, J. Struct. Div., ASCE, 104(11)

pp. 1763-1773(1978).

27.　Popovics, P., Stress-Strain Relations for Concrete Under Compression, Am. Concr. Inst. J., 61(9), pp. 1229-1235 (1964).

28.　Saenz, L. P., Discussion of Equation for the Stress-Strain Curve of Concrete, by P. Desayi and S. Krishnam, ACI Journal, Proceedings, Vol. 61 No. 9 (1964).

29.　Pike, P. J., Buykozturk, O. and Connor, J. J., Thermomechanical Analysis of Refractory Concrete Lined Coal Gasification Vessels, Research Report No. R80-2, Dept. of Civil Engr., MIT, U.S. Dept. of Energy, Jan., 1980.

7
Creep Data

I. INTRODUCTION

The term *creep* has often been used in the literature to identify the time-dependent straining of refractories or other materials in general. There are two tests to evaluate or quantify the time dependent straining. They are the *creep test* and the *relaxation test*. The creep test is basically a time-dependent stress-controlled loading test, while the relaxation test is a time-dependent strain-controlled loading test. More details regarding these two tests will be discussed later in this chapter. However, for the purposes of the following discussion, the term creep will be used, unless otherwise specified, as a term to identify the time-dependent straining of refractories.

Creep of refractories is another mechanical material property that plays an important role in understanding the structural behavior of the refractory lining system. The previously discussed compressive stress-strain data are used in the structural analysis to evaluate the instantaneous analysis or time-independent behavior of the refractory structure.

II. BACKGROUND

Refractory creep increases at high temperatures. The creep threshold temperature, or the temperature at which refractory creep should be considered, varies among the various types of refractories. In many cases, refractory linings, such as those used in cylindrical vessels, are exposed to a through-thickness temperature gradient. Only a small region on the hotface side of the lining is exposed to temperatures at which creep is significant. If one is evaluating the total thermal expansion interaction between the refractory lining and support structure, neglecting this small hotface region of creep may not result in a significant error in the solution. The reason is that this region has already been significantly deformed by the instantaneous plastic straining. Also, this region is typically very small and the resulting total expansion force in this hotface region is quite small compared to the remainder of the lining expansion force. Also, if one is evaluating the greatest expansion force interaction between the lining and support structure, the greatest forces will be with the instantaneous analysis results. Inclusion of the creep behavior will result in a relaxation of the hotface refractory material and a lesser expansion force. Again, however, the degree of the total reduction of the thermal expansion force can be quite small when creep is considered. If the investigative effort is to evaluate the stress-strain behavior of the local hotface region of the lining system, then the creep behavior of the refractory lining material should be included in the structural analysis.

One additional consideration should be made in using the available refractory creep data. As previously discussed, most of the creep data have been developed at low compressive stress levels, considerably less than the compressive stress levels experienced in actual lining systems. There is a concern over applying these creep data to lining refractories that are developing significantly higher compressive expansion stress.

Just as with the development of the compressive stress-strain data presented in Chapter 6, the creep data were developed by several testing laboratories. The laboratories include refractory companies, independent laboratories and universities. Considerably more creep studies have been conducted than compressive stress-strain studies.

The objective of this chapter is not to rate or rank the refractory materials presented with regard to strength or other aspects associated with the refractory life, but rather to provide insight into those interested in creep deformations that occur in refractories in order to aid in designing better refractory lining systems.

III. CREEP RESPONSE OF A REFRACTORY LINING STRUCTURE WHEN SUBJECTED TO STRESS-CONTROLLED AND STRAIN-CONTROLLED LOADINGS

Typical examples were previously provided with regard to the use of compressive stress-strain data associated with stress-controlled and strain-controlled loadings. The following simplified examples will demonstrate the influence of creep on a refractory lining subjected to stress-controlled and strain-controlled loadings. The examples used here are similar to those used in Chapter 5. However, they serve to show the role of the load type on the creep response of the refractory structure.

The creep response of refractory materials has been expressed in different equation forms. One popular form [1-4] of the creep equation, defining the total percent of creep strain (ϵ_c), is:

$$\epsilon_c = e^a \, t^b \, \sigma^c \, e^{-Q/PT} \tag{7.1}$$

where a, b and c are constants unique to the refractory material under consideration, e is the constant 2.71828 . . ., t is the time period of interest, σ is the applied stress, Q is the activation energy, P is the gas constant and T is the absolute temperature.

Another popular form of the creep equation is [5-10]:

$$\epsilon_c = e^a \, t \, \sigma^c \, e^{-Q/PT} \tag{7.2}$$

or:

$$\epsilon_c = A f(s) \, t \, \sigma^c \, e^{-Q/PT} \tag{7.3}$$

where A and f(s) are constants unique to the material under consideration. f(s) is defined as a structure term.

Basically, all three equation forms are similar. The constant terms e^a, A and f(s) are constant terms that do not vary with time. The definition of time term in equation Equation 7.1, t^b, can have b = 1 in which the time is linear.

Figure 7.1 describes the total creep strain that would occur in a refractory material subjected to a constant load F. ϵ_o is the initial strain that occurs due to the time-independent displacement of the material. ϵ_o is a combination of the elastic and plastic deformation due to the applied load. This initial strain would be determined by the compressive stress-strain curve, as illustrated in Chapter 6. The initial strain is defined as:

$$\epsilon_o = F/AE \tag{7.4}$$

The creep strain is obtained by substituting the applied stress, the material constants and the elapsed time ($\Delta t = t_o\text{-}t_f$) into Equation 7.1. Note that the applied loading is constant for the elapsed time under consideration. The creep response of materials is sometimes expressed as two parts, primary (or initial) creep strain (ϵ_{pc}) followed by secondary creep strains (ϵ_{sc}). The primary creep and secondary creep strain are typically expressed by separate and distinct equations. Usually primary creep becomes negligible after a short time period into the time-dependent response of the material. This initial straining is represented by the initial non-linear curve of creep strain versus time. The secondary creep strain typically continues to increase linearly as a function of time. During the final stages of creep when the material begins to fail because of excessive creep, the creep response becomes more accelerated. This final stage of creep is called *tertiary creep*.

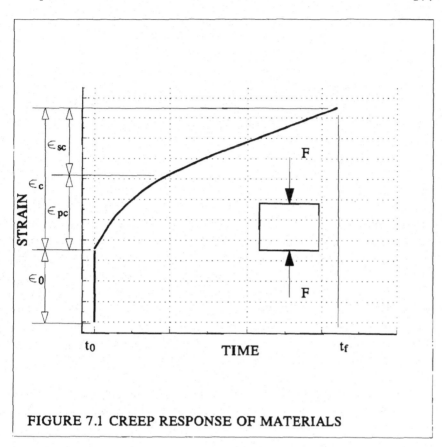

FIGURE 7.1 CREEP RESPONSE OF MATERIALS

In some cases investigators of refractory creep have shown the initial creep to be quite small and insignificant. Other investigators have shown the secondary creep to be non-linear. Our purpose here is not to interpret or question the creep data results, but rather to present the data of these investigations. If the refractory user desires to include the creep response of his refractory materials under consideration, it would be a prudent design procedure to request creep data for the materials under consideration from the refractory manufacturer.

A. Creep Response of a Refractory Lining System Subjected to a Stress Controlled Loading

The stress-controlled loading on a refractory lining structure, as previously discussed, implies that the internal equilibrating stress can be defined by using only the structure geometry. In the case of our simplified example, the geometry is the cross-sectional area A, illustrated in Figure 7.1.

The use of Figure 7.1, which illustrates the results of a creep test under constant load, is also used to illustrate the concept of a refractory lining system exposed to a stress-controlled loading because the creep test is a stress-controlled load test. In this example, to illustrate the creep effects of a stress-controlled load, the load is assumed to be constant during the time interval under consideration. This assumption is quite compatible with the concept of a gravity load (a stress-controlled loading) imposed on a refractory structure. It can be concluded that the creep strain continues to increase in time for a stress-controlled loading. As previously discussed, stress controlled loadings are typically quite small. Therefore, the total creep strain rate can also be small.

In refractory systems where large stacks of refractory shapes are used, creep analysis is conducted to evaluate the total creep subsidence due to gravity loading. Typically, the calculations are quite simple and similar in nature to those defined in the previous example. The refractory block weight is known as well as the time period under consideration. The refractory manufacturer can most likely define the creep equation for the material of interest. If the temperature environment is fairly constant with time, then the evaluation of the total creep strain is quite straightforward.

In summary, for a lining exposed to a stress-controlled loading, the creep strain will continue to increase in time. There are some exceptions, but typically the stress-controlled load is constant in time causing the creep strain to continuously increase in time.

Also, the stress-controlled loads are quite low which means that

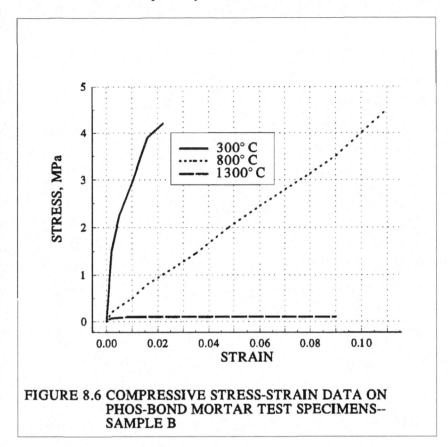

FIGURE 8.6 COMPRESSIVE STRESS-STRAIN DATA ON
PHOS-BOND MORTAR TEST SPECIMENS--
SAMPLE B

Figure 8.7 summarizes the calculated mortar MOE using the
Figure 8.6 data and Equation 8.5. The data points used to construct
curves are identified. Comparing Figures 8.3 and 8.7, there is a
favorable comparison between the two completely independent test
sample results. The mortar MOE at the low temperature range is
somewhat higher for Sample B than for Sample A. Within the
middle temperature range (800-1100°C), both samples are quite
similar. At the higher temperatures, there is a consistency with
respect to the influence of temperature. The Sample A upper
temperature is 1200°C while the Sample B upper temperature is
1300°C.

The calculated amount of deformation that occurs in the

the creep rate is quite low. It would take a significant amount of time to reach any significant total creep deformation.

If long time periods are considered in the creep analysis for the stress-controlled load condition, quite often only the secondary creep equation is used. When comparing the amount of primary creep deformations to the secondary creep deformations for long time periods (years of time), the primary creep deformations are often insignificant.

B. The Creep Response of a Refractory Lining System Subjected to a Strain-Controlled Loading

The strain-controlled loading on a refractory lining structure, as previously discussed, implies an imposed strain. As in the example of Chapter 2, this simplified example is assumed to be totally restrained. Also as in Chapter 2, it is assumed the lining is heated from an ambient temperature T_A to an operating temperature T_S within a very short time period. The instantaneous thermal strain (ϵ_R) imposed on the lining is:

$$\epsilon_R = \alpha(T_S - T_A) \qquad (7.5)$$

where α is the coefficient of thermal expansion. For the purposes of this discussion, it is assumed that the lining is heated to a temperature T_S in a rapid manner such that no significant creep occurs during heatup. This assumption is necessary in order to describe an idealized creep response of the lining. The idealized creep behavior serves to better define the actual creep response. The lining is then assumed to be held at the T_S temperature.

The creep response of a lining exposed to a strain-controlled loading differs considerably from that of the lining subjected to a stress-controlled loading. First, the magnitude of the instantaneous stress $(S_{RO}$, see Figure 7.2) is a function of the instantaneous compressive stress-strain data described as:

$$S_{RO} = E\epsilon_{RO} \qquad (7.6)$$

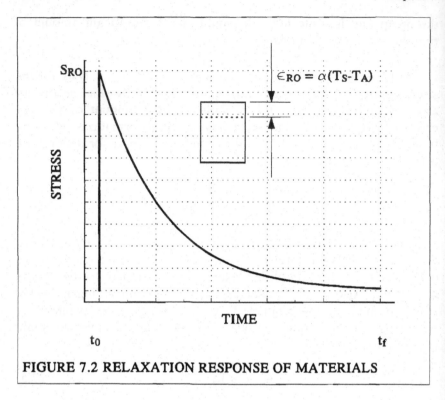

S_{RO}

$\epsilon_{RO} = \alpha(T_S - T_A)$

STRESS

TIME

t_0

t_f

FIGURE 7.2 RELAXATION RESPONSE OF MATERIALS

E is the modulus of elasticity, taken as the slope of the compressive stress-strain curve at the temperature T_s and for the thermal strain of ϵ_{RO} (see Equation 7.5).

As illustrated in Figure 7.2, the imposed strain is constant with respect to time, while the stress decays or relaxes with respect to time. The creep response of the lining with a strain-controlled loading is basically the reciprocal of the lining with a stress-controlled loading. It can be assumed that for an extremely long time period (or infinite time), the stress will decrease to a zero value and that the total creep strain (ϵ_c) will be equal to the imposed strain:

$$\epsilon_c = \epsilon_{RO} \tag{7.7}$$

The strain-controlled loading results in a reciprocal creep response. The solution for the constant strain material response is the reciprocal of the creep response. The dependent variable is stress and defined as:

$$\sigma_c = \epsilon_c (e^a e^{-QT/P} t)^{-1/c} \qquad (7.8)$$

The time-dependent response of materials is occasionally measured by a relaxation test. The relaxation test is the reciprocal of the creep test in that the strain is constant in the relaxation test, while the stress is constant in the creep test. In most cases, the creep test is a more economical and usually less complicated test to conduct. The relaxation creep test can also be defined as a strain-controlled loading test.

In summary, the refractory lining with a strain-controlled loading results in an upper limit of total creep strain equal to the initial imposed thermal strain. Also, for the strain-controlled loading, the stress will decrease in time and ultimately become zero when all of the imposed strain is relaxed by creep.

The strain-controlled loading on a refractory structure in combination with time-dependent straining results in a limited amount of creep strains within the lining system. For an assumed infinite time period, the maximum creep strain will be equal to the restrained portion of the imposed thermal strains.

IV. EXAMINING THE CREEP EQUATIONS

To better understand the creep equation in predicting creep strain, the various parts of the creep equation are examined here. Typical values are used for the various creep equation parameters. The creep equation parameters are defined here as a, b, c, A and f(s). The influence of the stress σ and temperature T on creep straining is also examined. The purpose here is to provide insight into sensitivity of each portion of the creep equation with regard to the amount of predicted creep strain.

It should be noted that the methodology of defining the time-independent stress-strain material properties (static compressive stress-strain data) and the time-dependent stress-strain material properties (creep data) is done in basically two different ways. The static compressive stress-strain data are typically presented in the form of data curves at selected temperatures. In some cases, an algebraic equation is used to define each static compressive stress-strain curve. This form of data presentation was done for some of the initial stress-strain data in Chapter 6. In most cases, however, the form used to present the static compressive stress-strain data is data curves.

Creep data, on the other hand, is in most cases presented in equation form in which the constants of the equation are presented. In some cases, the total creep deformations are presented as a result of the test results. However, in order to apply the creep data, the creep equation form and the creep equation constants must be defined, simply because the use of computer models requires the use of a creep equation. In the case of the static compressive stress-strain data, computer modeling in most cases uses a bilinear or multilinear formulation of the actual stress-strain curve.

A. The Stress Exponent c

The stress exponent c is examined here to evaluate the impact of this constant on the creep equation's prediction of creep strain.

Based on various investigators, the stress exponent c can vary from 0.3 to nearly 2.0. For the purpose of this discussion, the period is arbitrarily set at 8,500 hours, or approximately one year (354 days). The remainder of the creep parameters for the equation form of 7.1 were set at the following values [3]:

$$
\begin{aligned}
a &= 5.10 \\
b &= 0.40 \\
Q &= 34 \text{ kcal/mole} \\
P &= 0.0011036 \\
T &= 3260°R \ (2800°F)
\end{aligned}
$$

Note that the temperature T is in degrees Rankine (°R) in the creep equation. The stress is set at 5 psi (0.0345 MPa). The units defined in the following discussion on creep are those reported by the various investigators.

The results of using a range of c values is described in Figure 7.3. The value of c = 0.9 was used by the investigator [3]. The investigator's analysis was carried out to about 600 hours, in which a total creep deformation of 5% was achieved. This was replicated for the time of 600 hours shown in Figure 7.3. The total creep deformation of about 30% is reached in 8,500 hours for c = 0.9. The refractory would most likely fail at a much lower amount of creep deformation. When the c parameter is reduced to 0.3, a much lower creep deformation occurs (about 11% at 8,500 hours). The characteristics of the materials determine the value of c and the value of the other parameters of the material being tested. The purpose here is to reflect on the c value influence on creep. In most cases, the material with the least amount of creep is desired. Therefore, a refractory material with a low c value would be

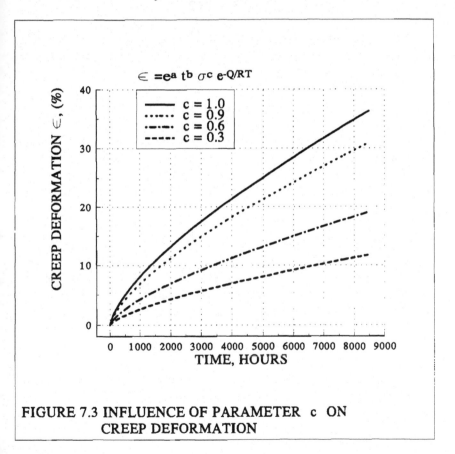

FIGURE 7.3 INFLUENCE OF PARAMETER c ON
CREEP DEFORMATION

chosen (assuming other parameters are equal), since a low c value results in less creep deformation.

For a constant stress, the c parameter does not influence the shape of the creep deformation curve. As shown in Figure 7.3, the c parameter is basically a parameter that controls the magnitude, or the influence of the stress, in causing creep deformation. For example with c = 1, the value of σ^c for this example is (using stress equal to 5 psi):

$$5.0^1 = 5.0$$

At the lower range of c = 0.3, the value of σ^c is:

$$5.0^{0.3} = 1.62$$

Therefore, the effectiveness of the applied stress is governed by the creep stress exponent c. For a stress of 5 psi, the exponent c of 0.3 reduces the effectiveness of the creep stress to:

$$(1.62/5)(100) = 32.4\%$$

of the full value of stress (c = 1.0). As the c exponent decreases, likewise the effectiveness of the creep stress to cause creep also decreases, assuming a constant value of stress. The c value can also be said to reflect the resistance of the material against creep deformation.

Figure 7.4 describes the impact of an increasing creep stress for two different values of the creep stress exponent c. At the stress level of 5 psi, the creep stress effectiveness is reduced to 32.4% in going from c = 1.0 to c = 0.3, as defined above. At a stress level of 2,000 psi, the effectiveness of the creep stress is reduced to:

$$(2000^{0.3}/2000^{1.0})(100) = 0.49\%$$

As the creep stress increases, the material becomes increasingly resistant to creep straining when compared to the full value of stress (c = 1.0).

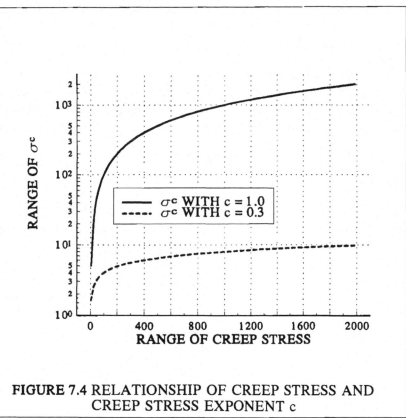

FIGURE 7.4 RELATIONSHIP OF CREEP STRESS AND
CREEP STRESS EXPONENT c

B. The Constants e^a and $Af(s)$

The constants e^a and $Af(s)$ affect the amount of creep deformation just as the stress exponent c does. When the constants e^a or $Af(s)$ are increased in value, the amount of creep deformation is increased in a direct proportional manner. This discussion addresses the relationship between the exponent a and the value of $Af(s)$. The constant e is the natural log base and is defined as 2.718281828 . . . .Various investigations define an a value that ranges from a minimum of 1.0 to a maximum value of about 45. Therefore, the constant a has a much greater range in magnitude than the stress exponent c.

As previously discussed, both the e^a and the Af(s) constants are similar in that they are independent of time. Figure 7.5 describes the relationship between a and e^a. Since e^a is equal to Af(s), the exponent a can also be related to the value Af(s). The material investigators may not agree that these constants are similar. But in terms of the mathematical usage of these constants, they are similar. Figure 7.5 shows a significant increase in the value of e^a when the exponent a is increased from a value of 4 up to a value of 11. That is:

$$e^4 = 54.60$$
$$e^{11} = 59,874.14$$

Therefore, the a constant has a profound influence on the magnitude of creep strain predicted by the creep equation. Figure 7.6 describes

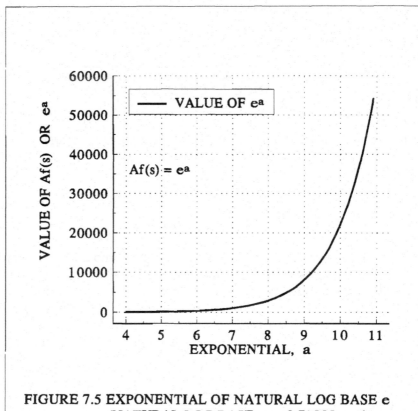

FIGURE 7.5 EXPONENTIAL OF NATURAL LOG BASE e
(NATURAL LOG BASE e = 2.71828 - - -)

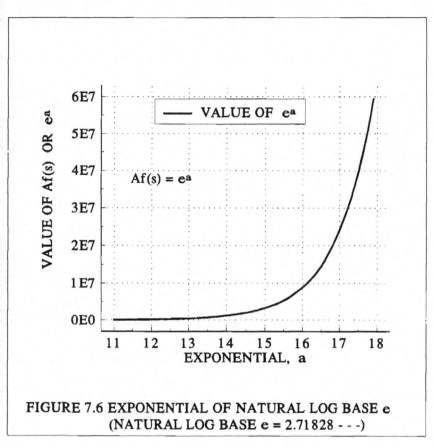

FIGURE 7.6 EXPONENTIAL OF NATURAL LOG BASE e
(NATURAL LOG BASE e = 2.71828 - - -)

the value of e^a for the range of 11 to 18 over a constant a. The value of e^a for a value of a equal to 11 indicates a value of about 6×10^7. This is similar in magnitude to the value of Af(s) reported by investigators of creep. The plotting of the relationship between a and e^a using a linear scale is somewhat cumbersome as illustrated in figures 7.5 and 7.6. Using a log scale for e^a values provides more insight into the relationship between a and e^a, as shown in Figure 7.7. Figure 7.7 provides additional insight into the impact of the exponent a in the prediction of creep strain.

C. The Time Exponent Constant b

The time exponent constant b can vary from minimum values

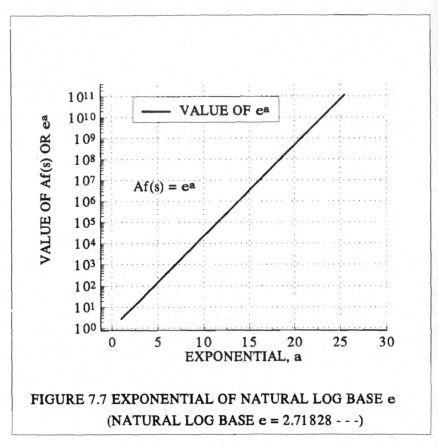

FIGURE 7.7 EXPONENTIAL OF NATURAL LOG BASE e
(NATURAL LOG BASE e = 2.71828 - - -)

of 0.4 to a maximum value of 1.0 according to investigators.

Therefore, when comparing the range of the exponents a, b and c, the time exponent b has the least range. Just as with the stress exponent, the lower the value of the exponent b, the less creep strain there is for a given range of time, all other creep parameters being equal.

Figure 7.8 describes the influence of the time exponent b in predicting the creep strain. As shown here, the lower value of b reflects a *time-hardening* effect. That is, as this exponent b decreases, the material is less susceptible to the effect of time in causing creep. Figure 7.8 is for the time span of 0 to 250 hours.

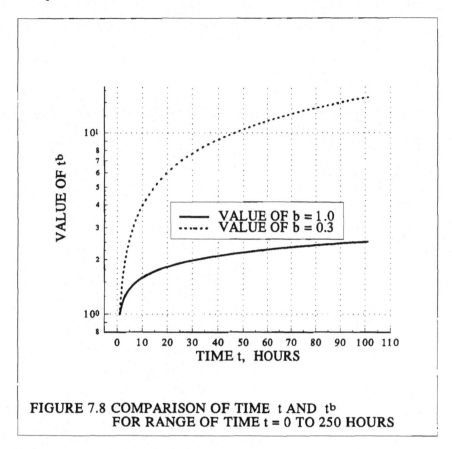

FIGURE 7.8 COMPARISON OF TIME t AND t^b
FOR RANGE OF TIME t = 0 TO 250 HOURS

Figure 7.9 reflects the effect of time for the time range of 0 to 8,750 hours (about 1 year). The influence of the exponent b can be expressed for the two extreme values for b: 1.0 and 0.30. At 250 hours:

$$(250^{0.30}/250^{1.0})(100) \; = \; 2.10\%$$

At 8,750 hours:

$$(8,750^{0.30}/8,750^{1.0})(100) \; = \; 0.17\%$$

Therefore, as time increases, the lower-valued b exponent has a more significant influence in reducing the creep strain. It can also

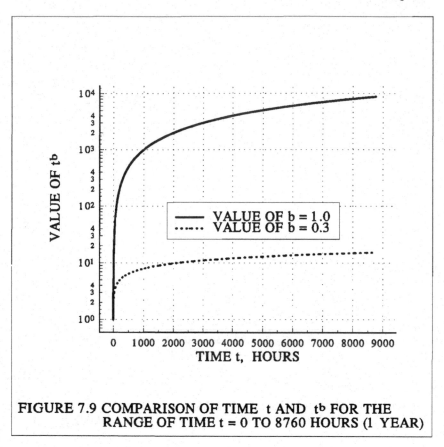

FIGURE 7.9 COMPARISON OF TIME t AND t^b FOR THE
RANGE OF TIME t = 0 TO 8760 HOURS (1 YEAR)

be concluded that the time-hardening effect in reducing creep
strain becomes more pronounced.

D. The Influence of Temperature

The last element of the creep equation that will be examined
is temperature.   The temperature is typically expressed as an
absolute temperature in degrees Rankine (°F + 460) or Kelvin (°C
+ 273). The temperature term of the creep equation, $e^{-Q/PT}$, has the
absolute temperature in the denominator of the e exponent.   The e
constant is the natural log base, just as for the leading term e^a in the
creep equation.   The Q constant is the activation energy and the P
constant is the universal gas constant.   The universal gas constant

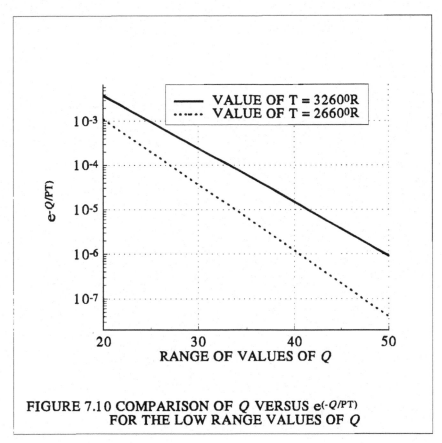

FIGURE 7.10 COMPARISON OF Q VERSUS $e^{(-Q/PT)}$
FOR THE LOW RANGE VALUES OF Q

can be expressed in many ways, as illustrated in Table I. The units of the activation energy Q, the universal gas constant P, and the absolute temperature T must be consistent.

Materials investigators have documented values of Q ranging from 19 to about 140 kcal/mole for the various types of refractories. The impact on the Q value, assuming a constant temperature 3260°R (2800°F), is examined in Figures 7.10 and 7.11. The value of R, associated with the units for Q, is 0.0011036 kcal/mole K. As the value of Q increases, the magnitude of $e^{-Q/PT}$ decreases. Also, as the value of Q increases, a greater difference exists between the two temperatures chosen in this example. The lower the temperature, the greater the decrease in $e^{-Q/PT}$ with the increasing

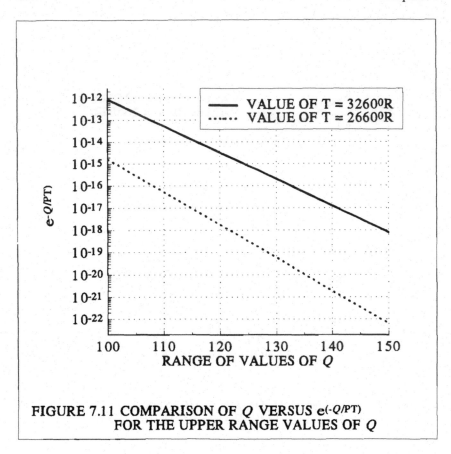

FIGURE 7.11 COMPARISON OF Q VERSUS $e^{(-Q/PT)}$
FOR THE UPPER RANGE VALUES OF Q

Q value. That is, as the value of Q increases, the amount of creep decreases. As the temperature decreases, the creep decreases. The later observation is, most likely, intuitive and obvious for most users of refractories and other materials.

E. Summary

Hopefully, the examination of the creep equation parameters has provided more insight into their influence on the amount of creep. If creep data are to be used, they should be verified by testing the refractory in question or at least substantiated in some scientific manner.

Table I

Universal Gas Constant P

VALUE	UNITS
1.987	BTU/lb mole °R
1.987	Cal/mole °K*
0.0011036	k cal/mole °R*
8.315	kJ/mole °K*
8.315×10^7	ergs/mole °K
1545	ft lb/lb mole °R
53.3	ft lb/lb°R
287	J/kg°K
8.205×10^{-2}	m^3atm/kmole °K
8.314×10^{-2}	m^3bar/kmole °K

*Units typically used in literature.

V. DEFINITION OF CREEP CONSTANTS

The previous sections defined the influence of the various creep equation constants as applied to the creep equations shown in Equations 7.1, 7.2 and 7.3.

A summary of refractory creep investigators is provided in Table II. In most cases the refractories have been tested at very low stress levels. In one instance the creep test was conducted at 2,000 psi (14 MPa). In most instances, the low level creep data did not exhibit significant creep straining at temperatures below about 1100°C (2000°F). However, the use of this low stress level creep data for higher stress applications is questionable.

TABLE II

Creep Equation Parameters
for Creep Equations in the Form of Equations 7.1, 7.2, and 7.3

Type of Refractory	Creep Equation Constants					Test Stress	Units	Ref. No.
	$Af(s)$ $\times 10^n$	b	c	Q $\times 10^n$	T			
Fireclay, superduty	1.12 (12)	1	1	1.19 (5)	K	0 to 100 psi	Q=cal/mole R=cal/K = 1.978 ϵ_c = %/hr. t = hr	6
60% Al_2O_3 Brick	7.83 (5)	1	1	7.2 (4)				
65% Al_2O_3 Brick	6.83 (8)	1	1	9.6 (4)				
Mullite Brick	3.92 (8)	1	1	9.3 (4)				
Fireclay Brick	3.0 (9)	1	1	4.25 * (4)	R	28 psi	ϵ_c=in./in./hr	**
70% Al_2O_3 Brick	2.32 (9)	1	1	5.82 * (4)				
70% Al_2O_3 Brick	2.32 (9)	1	1	5.78 * (4)				
60% Al_2O_3 Brick	1.01 (5)	1	1	4.00 * (4)				
90% Al_2O_3 Castable	8.55 (-5)	1	1.68	4.10 (1)	R	3,000 psi	Q=kcal/mole	7
90% Al_2O_3 Castable	3.34 (14)	1	1	1.47 (2)				
90% Al_2O_3 Castable	2.42 (-11)	1	1.7	1.71 (2)	K	13.8 MPa	Q=kJ/mole	8
Mag Cr Brick,D.B.	2.5 (5)	0.80	1	3.5 (5)	K	2-4 MN/m^2	Q=kJ/mole R=8.315kJ/ mole K	5

Note: Exponent n for the Af(s) and Q terms shown in parentheses. Units for temperature in Rankine (R) or Kelvin (K).
* Constants shown for ratio of Q/R.
** Personal notes.

TABLE II (Continued)

Creep Equation Parameters
for Creep Equations in the Form of Equations 7.1, 7.2 and 7.3

Type of Refractory	Creep Equation Constants					Test Stress	Units	Ref. No.
	a	b	c	Q	T			
60% MgO Brick,D.B.	7.80	0.65	0.85	49	R	20 psi	Q=kcal/mole	4
MgO Brick,D.B.	-4.60	0.60	1.40	19			R=0.0011036 kcal/mole R	
MgO Brick,D.B.	2.10	0.55	1.30	34			ϵ_c= %	
MgO Brick,C.B.	22.10	0.30	0.90	81			t= hr	
50% Al_2O_3 Brick	8.8	0.50	0.75	41	R	28-57 psi	ϵ_c= % Q = kcal/mole	1
70%Al_2O_3 Brick	34.20	0.65	1.15	132		14-57 psi	t= hr R= 0.0011036	
80%Al_2O_3 Brick	22.15	0.80	0.85	91		28-57 psi	kcal/mole R	
93%MgO Brick,D.B.	15.65	0.40	0.30	71	R	14 to 57 psi	Q=kcal/mole	2
98%MgO Brick,C.B.	1.30	0.55	0.35	22			R=0.0011036 kcal/mole R	
95%MgO Brick,D.B.	5.10	0.70	0.90	34			ϵ_c= %	
92%MgO Brick,D.B.	5.30	0.65	0.60	31			t= hr	
50%MgO Brick,D.B.	10.85	0.55	0.75	57				
44%MgO Brick,D.B.	12.85	0.60	0.95	61				
21%MgO Brick,C.B.	41.45	0.55	0.80	141				

VI. OTHER FORMS OF THE CREEP EQUATION

Other forms of refractory creep equations have been developed to quantify creep [12-14]. They are:

$$\epsilon = At^m$$

where ϵ is the total creep strain, A is a function of stress (σ) and temperature (T), t is time and m is the time exponent. The creep rate is defined as:

$$d\epsilon/dt = mA^{1/m}\epsilon^{(1 - 1/m)}$$

in which the creep rate is a function of the creep strain.

Another form of a refractory creep equation is [14]:

$$\epsilon = \epsilon_{ss}T + \epsilon_p[1-e^{(-T/JK)}]$$

where ϵ_{ss} is the steady-state (or secondary) creep strain rate, ϵ_p is the total primary creep strain, and JK is an empirical time constant.

VII. SUMMARY

The time-dependent response of a refractory material can be measured by two test methods: the creep test and the relaxation test. Typically, the creep test is the first choice among most investigators because of lesser cost and effort. Also, the form of the creep test equation, with the creep strain being the dependent variable, is compatible with the equation form used in most current structural computer programs.

With regard to load types and creep of refractory lining systems, the stress-controlled loading, in combination with time-dependent creep straining of refractories, results in an unlimited accumulation of creep strains within the lining system. Typically, however, the stress-controlled loads are small, resulting in small accumulations of creep strain.

The creep data provided in Table II were, in most cases, developed for very low stress levels, considerably less than the stress levels experienced in actual vessel linings. As a result the use of this data for vessel lining investigations is questionable. The low levels of creep stress used in the creep test implies a need for creep tests at stress levels normally encountered in vessel lining systems.

REFERENCES

1. Stett, M. A., National Refractories & Minerals Co., Letter to C. A. Schacht regarding creep data, January 29, 1988.

2. Gilbert, S. V., Creep Equation Model Used to Study Checker Subsidence, Glass Industry, September, 1985.

3. Stett, M. A., Measurement of Properties for Use with Finite Element Analysis Modeling, American Ceramic Society 87th Annual Meeting, Trans. Cer. Bull., Vol. 64, No. 3, p. 484 (1985).

4. Neal, J. M., Kaiser Refractories, Letter to C. A. Schacht regarding creep data, October 23, 1984.

5. Dixon-Stubbs, P. J. and Wilshire, B., High Temperature Creep Behavior of a Fired Magnesia Refractory, Transactions of British Ceramic Society, Vol. 80, pp. 180-185 (1981).

6. Ainsworth, J. H. and Kaniuk, J. A., Creep of Refractories in High Temperature Blast Furnace Stoves, American Ceramic Society Bulletin, Vol. 57, No. 7, pp. 657-659 (1978).

7. McGee, T. D. and Smyth, J. R., Creep Behavior of Monolithic Refractory Materials, Fourth Annual Conference on Materials for Coal Conversion and Utilization, National Bureau of Standards, Gaithersburg, MD, October 9-11, 1979.

8. Bray, D. J., Smyth, J. R. and McGee, T. D., Creep of 90+% Al_2O_3 Refractory Concrete, American Ceramic Society

Bulletin, Vol. 59, No. 7, pp. 706-710 (1980).

9. Brady, D. S., Creep of Refractories: Mathematical Modeling, Amer. Cer. Soc. Proceedings, New Devel. in Mono. Ref., Vol. 13, pp. 69-80 (1985).

10. Hasselman, D. P. H., Approximate Theory of Thermal Stress Resistance of Brittle Ceramics Involving Creep, Amer. Cer. Soc., Journal, Vol. 50, No. 9, pp. 454-460 (1967)

11. Krause, Jr., R. F., Compressive Strength and Creep Behavior of a Magnesium Chromite Refractory, Ceram. Eng. Sci, Proc., 7 [1-2], pp. 220-228 (1986).

12. Sun, J., Effect of Coal Slag on the Microstructure and Creep Behavior of a Magnesium-Chromite Refractory, Amer. Cer. Soc., Cer. Bulletin, 67 [7], pp. 1201-1210 (1988).

13. Lundeen, B. E. and McGee, T. D., Primary Creep Analysis of a Commercial Low Cement Refractory Castable, Amer. Cer. Soc., Ceramic Trans., Vol. 4, pp. 312-324 (1989).

14. Brady, D. J., Creep of Refractories: Mathematical Modeling, Amer. Cer. Soc., Advances in Ceramics, Vol.13, pp. 69-79 (1985).

8
Thermomechanical Aspects
of Joints

I. INTRODUCTION

The joint compressive stress-strain behavior is an inherent part of defining the compressive stress-strain behavior of refractory shapes. The more joints the lining has, the greater the influence the joints will have on the lining behavior.

Joints may be either mortared joints or dry (unmortared) joints. In either case, the stress-strain behavior of the joint interface differs considerably from the stress-strain behavior of the parent refractory material.

The primary emphasis here will be on the compressive stress-strain behavior of the joint. The joint cannot not develop significant tensile load. For this reason, the joint will separate under tensile loading.

The joint is a non-linear structural mechanism of the refractory lining system. The compressive stress-strain behavior is non-linear, even at low temperatures. The joint is also non-linear with respect to the joint contact area. When the lining is installed all joints are in full contact, or at least construction and installation methods attempt to make all joints have full contact. However, when the lining production loading (operating pressure and heatup temperature) is displaced, some joints may partially open. More details on joint opening behavior are provided in Chapter 9.

Joints (either mortared or unmortared) cannot resist significant tensile loading. Since the compressive loading is the significant loading, the objective here is to provide the refractory user with information regarding the compressive stress-strain behavior of mortared and unmortared joints. With an understanding of the joint behavior, better longer-life lining systems can be designed by the user. With increased awareness of joint behavior, better lining designs can also be provided by the refractory manufacturer.

One of the beneficial aspects of joints is their expansion allowance behavior [1-5]. The joint tends to compress more than the parent material, effectively reducing the total stiffness of the lining system. This aspect of joints is especially helpful for refractory materials that have a high coefficient of thermal expansion and are also very stiff. One undesirable aspect of joints is that they allow deleterious chemicals of the process to penetrate the lining. These chemicals can accelerate the lining deterioration in the region of the joint.

II. BACKGROUND

The compressive behavior of a joint has a profound influence on the total structural behavior of a lining system. The test data on the compressive stress-strain behavior of both types of joints are limited to a few published test results. However, these results show that when subjected to compressive loading, the joint stress-strain curve is not linear. Also, the joint is considerably more flexible than the parent refractory shape. Additionally, the mortar joint compressibility does not increase in a linearly proportional manner with the joint thickness.

The joint test data described in the following discussions were obtained from investigations on refractory lining joint systems and from structural masonry mortar joint systems. A considerable amount of investigative work has been conducted on the thicker structural masonry mortar joint. These later test results provide significant insight into the behavior of mortar joints and assist in a better understanding of the refractory lining mortar joint.

III. REFRACTORY MORTAR JOINT FUNDAMENTALS

The most fundamental elements of a mortar joint are described in Figure 8.1. The fundamental parameters are the thickness of the brick (t_b), in which all bricks are assumed of equal thickness, the

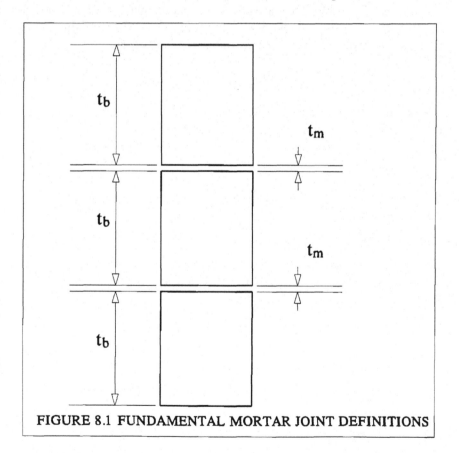

FIGURE 8.1 FUNDAMENTAL MORTAR JOINT DEFINITIONS

mortar joint thickness (t_m), the modulus of elasticity (MOE) of the brick (E_b), and the MOE of the mortar (E_m). The basic assumption is that the brick and mortar materials behave in a linear elastic manner.

A. Mortar Joint Properties

With the information as defined above, the MOE of the mortar can be determined using static compressive stress-strain data developed on core-drilled samples with a mortar joint. The mortar MOE is quite useful for evaluating the effective expansion forces of the lining with mortar joints. The following is a method of determining the MOE of the mortar.

Let us first introduce a very basic mathematical definition of the brick/mortar joint behavior. The compressibility of the brick when subjected to an arbitrary compressive load P, is:

$$dt_b = Pt_b/AE_b \tag{8.1}$$

where the cross-sectional area of both the brick and mortar is defined as A . Assuming a unit cross-sectional area:

$$dt_b = Pt_b/E_b \tag{8.2}$$

Similarly, the mortar joint will compress by amount Δt_m, defined as:

$$dt_m = Pt_m/E_m \tag{8.3}$$

The composite material with an MOE of E_{mb} is assumed to compress the same amount as the brick and mortar, or:

$$dt_{mb} = P(t_m + t'_b)E_{mb} \tag{8.4}$$

Typically, the test sample length for the composite system is the same as the brick sample length. Therefore,

$$t_b = t_m + t'_b$$

Setting Equation 8.4 equal to Equations 8.2 plus 8.3:

$$P_{tb}/E_{mb} = P_{tb}/E_b + P_{tm}/E_m$$

Canceling similar terms and rearranging to evaluate E_m:

$$E_m = E_{mb}E_b t_m/t_b(E_b - E_{mb}) \tag{8.5}$$

Often when static compressive tests are conducted on brick products, the tests use whole core-drilled brick samples and core-drilled split samples with a mortar joint of a defined thickness (t_m). With this data, all of the parameters of Equation 8.5 can be determined. The mortar material cannot be tested using core samples made of only mortar. The static compressive stress-strain data on all mortar samples usually underpredict the effective stiffness of the mortar material. The reason is that the mortar material is confined in the mortar joint, prohibiting inelastic flow of the mortar material when subjected to the compressive loading. The mortar constrained in the mortar joint is much stiffer, due to the joint restraint, than the stiffness values predicted using pure mortar samples.

Sample A Data

Static compressive stress-strain data on a 70% alumina brick and the 70% alumina brick with mortar joints are described in Figure 8.2 [5,6]. This mortar joint test data will be called Sample A data. The whole core-drilled samples of the 70% alumina brick are shown as the solid data curves. The mortar joint split core drilled samples are shown as the dashed data curves. All of the core samples were 64 mm (2.5 in.) in diameter. In both the whole and split mortar joint samples, the strain was measured over a 50-mm (2-in.) mid-length of the 102-mm (4-in.) long samples. In the mortar joint samples, the mortar joint was 1.59 mm (1/16 in.) thick. The mortar was a phos-bond type.

The following calculations are used to demonstrate the usefulness of the static compressive stress-strain data in evaluating the mortar MOE. The dashed curves represent the split core-drilled brick/mortar joint behavior for the dimensions defined. For the purposes of this demonstration, the mortar MOE is evaluated at the three temperatures of 20-815°C, 1000°C and 1200°C. The calcula-

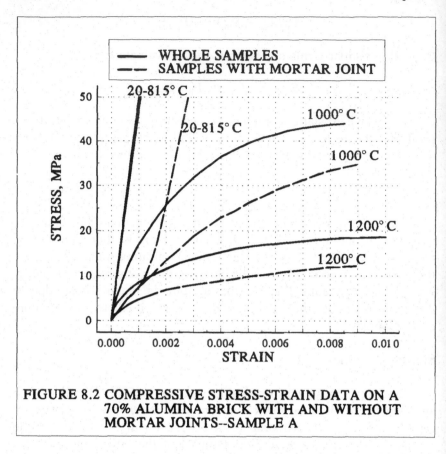

FIGURE 8.2 COMPRESSIVE STRESS-STRAIN DATA ON A
         70% ALUMINA BRICK WITH AND WITHOUT
         MORTAR JOINTS--SAMPLE A

tions are also conducted for various levels of stress for each
temperature. All three temperature sets of calculations are not
defined here. Rather, a typical set of calculations are for a defined
temperature and stress which is representative of how the mortar
MOE is evaluated.

The data curves for the 20-to-815°C temperature range are
nearly identical. It will be assumed that the stress-strain behavior
of the subject materials in Figure 8.2 are identical for this
temperature range. The following calculations are conducted for the
20-to-815°C temperature at a compressive stress of 10 MPa. In all
of the calculations, the secant modulus is used to define the MOE
for the given stress level (see Figure 3.2). For the whole

sample data, using Equation 8.5:

$$E_b = 10/0.00023 = 43.48 \text{ GPa}$$

The composite MOE is:

$$E_{mb} = 10/0.0012 = 8.3 \text{ GPa}$$

With t_b = 50 mm and t_m = 1.59 mm, solving for E_m:

$$E_m = (8.3 \times 43.48 \times 1.59)/50(43.48 - 8.3)$$

$$= 0.33 \text{ GPa}$$

These data imply that the mortar MOE is considerably softer than the brick MOE. The resulting MOE data for the range of stresses and temperatures considered is summarized in Table I.

Table I

Calculated Secant MOE Using Figure 8.2
Data and Equation 8.5

Data Temperature, C	Selected Stress, MPa	Secant MOE, GPa		
		E_b	E_{mb}	E_m
20-815	10	43.48	8.3	0.33
	25	46.30	12.76	0.56
	50	47.17	17.86	0.92
1100	10	18.52	7.14	0.32
	20	14.29	6.06	0.30
	30	10.91	4.62	0.24
1200	4	13.33	5.00	0.21
	8	9.41	2.67	0.11
	12	5.22	1.46	0.070

The calculated mortar joint MOE (from Equation 8.5) is plotted in Figure 8.3. The data points used for each curve are identified. Of particular interest is the stress-versus-MOE relationship of the mortar joint. These data indicate that the MOE is a function of the compressive stress applied to the joint. For the lower temperature range (20-815°C), the mortar joint MOE increases in a nearly linear manner with respect to an increasing applied compressive stress. This trend seems intuitively correct since the mortar is expected to densify with increasing stress. At the higher temperatures (1100 and 1200°C), the mortar MOE decreases in a non-linear manner with increasing applied compressive stress.

The initial response to the 1100 and 1200°C temperature mortar MOE data does not appear intuitively reasonable. As an

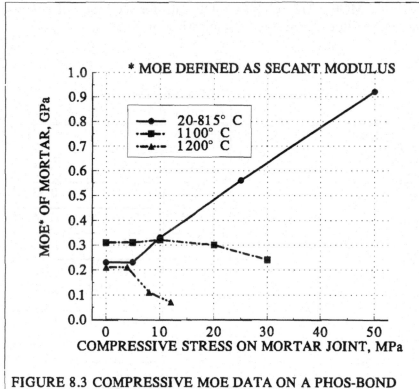

FIGURE 8.3 COMPRESSIVE MOE DATA ON A PHOS-BOND
            MORTAR USED IN A 1.59-mm THICK MORTAR
            JOINT--SAMPLE A

increasing load is applied to the porous mortar, it would be reasonable to assume that the mortar would densify and cause the MOE to increase. According to these data at higher stress levels, the grain structure or the bonding system is apparently deteriorating.

The 1200°C data show the rate of decreasing MOE is lessening as the compressive load increases. These data appear correct. That is, the load increases and densification occurs, resulting in an increase of the MOE. The MOE calculation for very small stress levels becomes quite difficult since the strain level is quite small. Therefore, the zero stress level is assumed to be equal to the MOE at the smallest reasonable stress level. This is a reasonable assumption since small stress levels are not of significant consequence.

The amount of mortar joint deformation is determined by using the difference in the amount of strain between the whole and mortar joint samples. For a given magnitude of mortar joint loading, the joint deformation dt_m is:

$$dt_m = dt_{mb} - dt_b \tag{8.6}$$

The mortar joint deformation can also be expressed as a percent of the original unloaded mortar joint thickness defined as:

$$dt_m\% = [dt_mb - dt_b)/t_m] \times 100 \tag{8.7}$$

The deformations are determined using the strain data and sample length. Equation 8.7 is expressed as:

$$dt_m\% = [t_b(\epsilon_{mb} - \epsilon e_b)/t_m] \times 100 \tag{8.8}$$

where ϵ_{mb} and ϵ_b are the strains that occur in the mortar joint and whole samples, respectively. Since t_b is equal to t_{mb}, t_b is used.

Returning to Figure 8.2, the 20-815°C data curves and using the stress level of 10 MPa, the $\Delta t_m\%$ is calculated as:

$$dt_m\% = [50(0.0012 - 0.0002)/1.59] \times 100$$

$$= 3.14\%$$

Therefore, the 1.59-mm thick mortar joint compresses only 3.14 percent at a compressive load of 10 MPa. That is, the joint thickness after the load of 10 MPa is applied would be 1.59(1 - 0.0314) or 1.54 mm. Calculations are conducted as before for a range of stress levels for each of the three sets of data (20-815°C, 110°C and 1200°C) and plotted in Figure 8.4.

The mortar joint of deformation behavior is described in Figure 8.4. This calculated data is shown in Table II. The plotted data describe some interesting deformation characteristics of the mortar joint. At the lower temperature range of 20 to 815°C, the deforma-

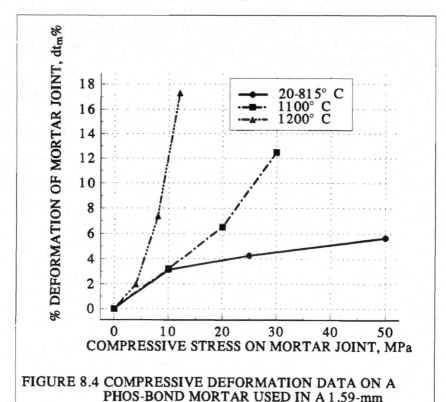

FIGURE 8.4 COMPRESSIVE DEFORMATION DATA ON A
         PHOS-BOND MORTAR USED IN A 1.59-mm
         THICK MORTAR JOINT--SAMPLE A

tion is linear with respect to the applied compressive stress, except near the origin. The non-linear behavior from the origin (zero stress) up to the 10-MPa stress level is attributed to an initial seating phenomenon. That is, seating is occurring between each end point of the sample and the testing machine platens.

At 1100°C and 1200°C the mortar joint deformation (as described in Figure 8.4) is non-linear. The rate of mortar joint deformation continuously increases with respect to the increasing the increasing applied compressive load. Also as shown, the increase in the mortar joint deformation increases with respect to an increasing temperature.

Sample B Data

A second set of stress-strain data regarding mortar joints were

Table II

Calculated Mortar Joint Percent Deformations
Using Figure 8.2 Data and Equation 8.7

Data Temperature,°C	Stress, MPa	%Δt_m
20-815	10	3.14
	25	4.25
	50	5.66
1100	10	3.20
	20	6.51
	30	12.48
1200	4	1.95
	8	7.39
	12	17.30

conducted on, what we will call Sample B. Figure 8.5 describes the static compressive stress-strain on whole brick specimens (in this case 80% alumina brick) and the split mortar joint specimens [7]. In this case the mortar joint was 2 mm thick and the test prisms were 30 mm by 30 mm in cross section. This mortar is also a phos-bond type mortar as was Sample A. The strain was measured over a 50 mm sample length.

Sample B specimens were also subjected to static compressive stress-strain tests as shown in Figure 8.6. As noted, the magnitude of the compressive stress was quite limited due to the low unconfined strength of the mortar samples. The MOE of the pure mortar samples will be compared later to the calculated mortar joint MOE.

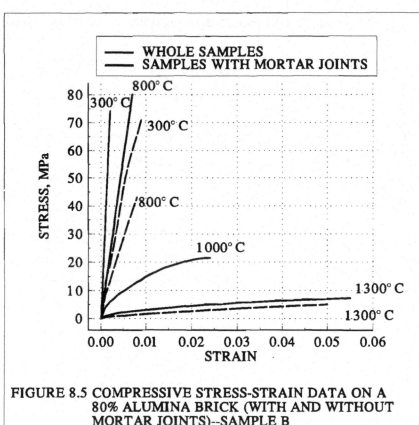

FIGURE 8.5 COMPRESSIVE STRESS-STRAIN DATA ON A
          80% ALUMINA BRICK (WITH AND WITHOUT
          MORTAR JOINTS)--SAMPLE B

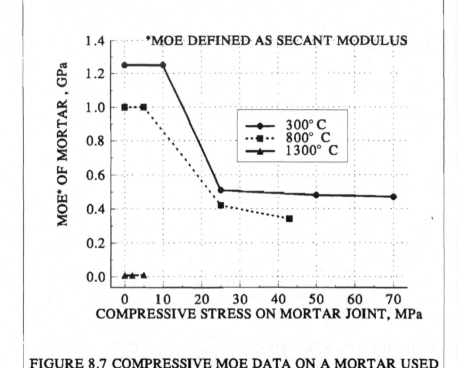

FIGURE 8.7 COMPRESSIVE MOE DATA ON A MORTAR USED
IN A 2-mm THICK MORTAR JOINT--SAMPLE B

Sample B 2-mm thick mortar joint is summarized in Figure 8.8. Sample B is evaluated at higher loads for the lower temperatures than Sample A, resulting in greater Sample B deformations at the higher loads. However, Sample A and B deformations are similar at these lower temperatures. At the medium temperatures, the deformations are also somewhat similar. At 1300°C, Sample B exhibits a considerable amount of deformation, approaching 55% at a loading of 5 MPa. This is expected because of the higher Sample B test temperature. Sample A, at 1200°C, undergoes about 17% deformation at a loading of about 12 MPa. The stiffer Sample A mortar at the 1200°C temperature may, in fact, be attributed to the temperature difference of 1200°C for Sample A and 1300°C for Sample B.

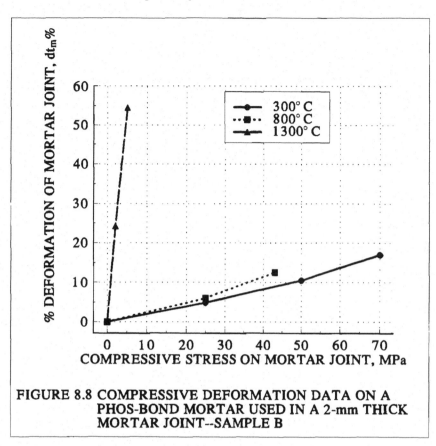

FIGURE 8.8 COMPRESSIVE DEFORMATION DATA ON A
PHOS-BOND MORTAR USED IN A 2-mm THICK
MORTAR JOINT--SAMPLE B

The MOE plot of the Sample B mortar test specimens is plotted in Figure 8.9. Because of the unconfined condition of the mortar specimens, the compressive loading did not exceed 5 MPa for any of the static compressive stress-strain data curves (see Figure 8.6).

B. Summary

The fundamental aspects of mortar joint behavior are the stiffness of the mortar joint as expressed by the MOE data and the deformation characteristics of the mortar joint. These data were developed using basic linear algebraic equations.

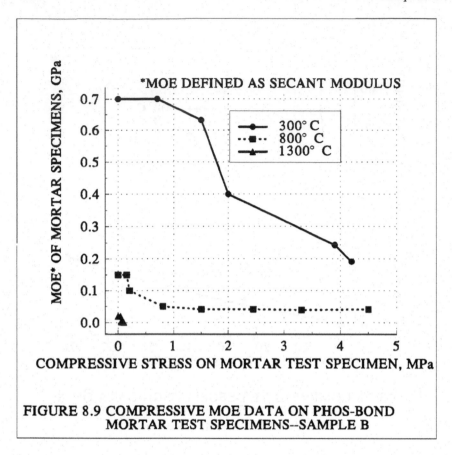

FIGURE 8.9 COMPRESSIVE MOE DATA ON PHOS-BOND
MORTAR TEST SPECIMENS--SAMPLE B

The resulting MOE calculations indicate that the MOE of the mortar (as constrained within the mortar joint) is highly non-linear. The mortar joint MOE is a function of the applied compressive stress and temperature.

The mortar within the joint appears to undergo a significant amount of densification and possible grain crushing. This is evident by the nature of the calculated MOE data curves. If only densification occurs, then the MOE should increase with increasing stress. In some cases, however, the MOE decreases with increasing stress. However, the rate of decreasing MOE appears to also decrease. This may imply that the effects of densification are playing a greater role than the grain crushing with increasing stress.

IV. FINITE ELEMENT ANALYSIS OF A MORTAR JOINT

In order to provide additional insight into the thermomechanical behavior, a finite element analysis was conducted with regard to compressive stress-strain behavior of the Sample B mortar joints [6]. The choice of the Sample B mortar joint data was based on the compressive strain data of the mortar specimens. The mortar data showed it to be considerably weaker than the same mortar contained in the mortar joints. The analysis replicated the compressive stress-strain tests on the split specimens with the mortar joint. However, the mortar material data used were those observed from the mortar specimen tests.

The dimensions of the Sample B split mortar joint specimen are described in Figure 8.10. The Y axis is in the vertical or axial direction, the direction of the loading. The X direction is the horizontal direction. Because of symmetry of loading and geometry, only a quarter section of the specimen contained within the X-Y coordinate axes was used in the analysis. As an additional simplification, the analytical model was assumed to be axially symmetric.

A. Analysis Results

The model used in the analysis is described in Figure 8.11. The bottom three rows of elements represent the half-thickness (1 mm) of the mortar joint. The remainder of the model is the 80% alumina brick. The analysis was conducted for the 800°C compressive stress-strain mortar data curve (see Figure 8.6). The mortar was assumed to be elastic for a very small stress range. The yield point was set at 0.20 MPa (30 psi). The elastic modulus was set at 100 MPa. The slope of the inelastic part of the data curve, as shown in Figure 8.6, was set at 44 MPa. In this analysis, it was assumed that the mortar would flow inelastically, as defined by the 800°C data curve. The compressive loading was increased from zero up to a maximum of 62 MPa (9,000 psi), close to the compressive loading used in the mortar joint tests (see Figure 8.5, 800°C data curve). The pressure was applied in a uniform manner across the top surface of the model (see Figure 8.11).

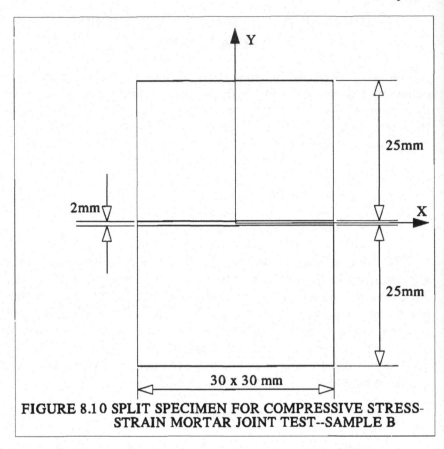

FIGURE 8.10 SPLIT SPECIMEN FOR COMPRESSIVE STRESS-STRAIN MORTAR JOINT TEST--SAMPLE B

The resulting stresses in the specimen, at the 9,000 psi loading, are shown in Figure 8.12. Because of the lack of restraint against plastic flow in the outer portion of the mortar joint, the mortar extrudes from the joint. The details of stresses in the outer region of the joint are shown in Figure 8.13. The minimum compressive stress* (MX) is in mortar that extrudes from the joint. The highest compressive stress (MN) is in the brick just above the mortar extrusion. A more detailed description of the mortar extrusion is shown in Figure 8.14. It should be noted that the extrusion is not to scale. All of the model displacements were amplified. However, the mortar joint displacements show the greatest amplitude.

---

* Using the rules of an algebraic scale, - is minimum,+ is maximum.

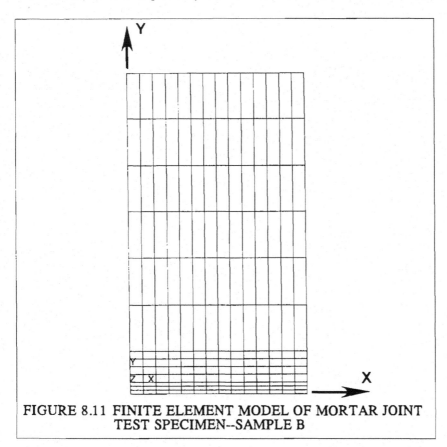

FIGURE 8.11 FINITE ELEMENT MODEL OF MORTAR JOINT
TEST SPECIMEN--SAMPLE B

Returning to Figure 8.12, the result of the mortar plastic shows lack of vertical support in the outer regions of the joints. However, at the interior portion of the joint (at model center, X = 0) the mortar was restrained from plastic flow, as evidenced by the greater vertical support and greater Y stress in this region. Except for the very local high compressive stress at the exterior portion of the specimen above the joint, the highest compressive stress occurs at the center of the specimen.

The analytical results also indicate that the joint will compress by an amount of 0.80 mm (for the half-joint thickness). The percent deformation is:

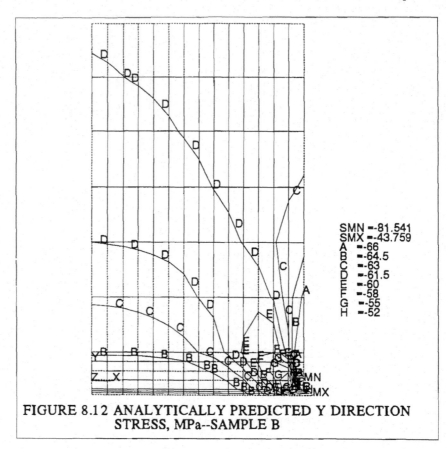

SMN =-81.541
SMX =-43.759
A =-66
B =-64.5
C =-63
D =-61.5
E =-60
F =-58
G =-55
H =-52

FIGURE 8.12 ANALYTICALLY PREDICTED Y DIRECTION
STRESS, MPa--SAMPLE B

$$\%t_m = (0.80/1) \times 100 = 80\%$$

Comparing this result to the mortar joint specimen deformation
results (see Figure 8.8, 800°C curve), this is quite excessive. The
actual data in Figure 8.8 show that a deformation of about 15%
should occur at a loading of 60 MPa at 800°C.

B. Summary

The analysis results verify that the mortar specimen
compressive stress-strain data are not representative of the actual
thermomechanical behavior of the mortar joint. They also verify
that the thermomechanical behavior of the mortar can only be

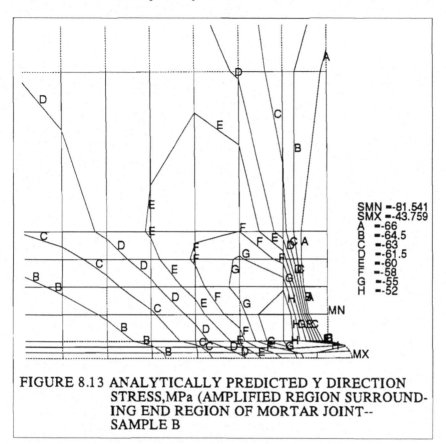

FIGURE 8.13 ANALYTICALLY PREDICTED Y DIRECTION
STRESS,MPa (AMPLIFIED REGION SURROUND-
ING END REGION OF MORTAR JOINT--
SAMPLE B

evaluated using mortar joint test specimens.

The results of both the analytical and test results imply that
considerably more research is needed to better understand mortar
joint behavior. Based on the available data and investigative work,
the following conclusions are made regarding mortar joint behavior:

a.   The compressive loading on the joint causes the
     mortar material to densify.

b.   The mortar joint MOE should increase as
     densification increases. All of the tests data does
     not support this assumption.

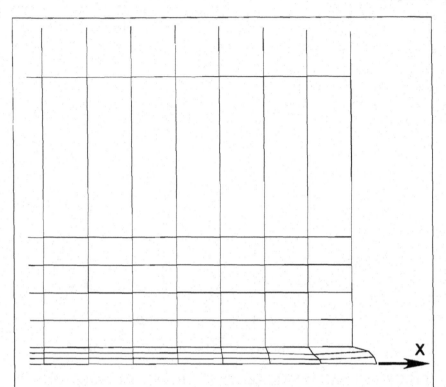

FIGURE 8.14 ANALYTICALLY PREDICTED DISPLACEMENTS
IN THE AMPLIFIED REGION SURROUNDING THE
END REGION OF MORTAR JOINT--SAMPLE B

c.   The mortar joint material grain structure appar-
     ently crushes when exposed to the higher
     compressive loads. This may be part of the
     densification procedure.

d.   The central portion of the mortar joint appears to
     resist more of the applied loading than the end
     portions of the mortar joints. This is attributed to
     the plastic flow of the mortar and the flow
     restraints of the mortar at the interior part of the
     mortar joint and the lack of flow restraint at the
     mortar joint ends.

e.   The use of linear equations predict that the MOE of the mortar is most likely an average MOE for the load range considered.

V. BEHAVIOR OF THE STRUCTURAL MASONRY MORTAR JOINT

The following discussion concerns the structural masonry mortar joint.   Masonry mortar joint is typically subjected to compressive and tensile loadings.   However, the masonry mortar joint is intended primarily for compressive loading, just like the refractory mortar joint. The only drawback in the masonry mortar joint investigative work is the lack of the thermal influence. Of course, masonry structures are subjected and exposed to only mild environmental temperatures. As a result, the thermal effects are not typically a part of masonry investigative work. The masonry mortar joint is typically much thicker than the refractory mortar joint. The maximum thickness of the structural masonry joint can be in the range of 10 to 13 mm while the maximum thickness of most refractory mortar joints is only a few millimeters.   In most high-temperature furnaces the refractory mortar joint is kept at minimum by dipping the brick in a thin mortar slurry during installation. Still the investigative work on structural mortar joints does provide insight into the behavior of refractory mortar joints.

A. Investigative Study on Structural Masonry Mortar Joints

A most comprehensive investigation on masonry mortar joints was conducted by McNary and Abrams [8]. Portions of their work which relate to the refractory mortar joint are reviewed in the following discussion.

The reason for the interest in the mortar joint was put best by McNary and Abrams, "Clay-unit masonry is composed of two materials with quite different properties; relatively soft cement-lime mortar and stiff fired-clay brick." The same is true for the refractory mortar joint system.

When the brick-mortar joint system is subjected to a uniaxial loading, the mortar has a tendency to expand laterally more than the brick. This mortar behavior was observed in the refractory mortar joint analytically predicted mortar deformations (see Figure 8.14). In the case of structural mortar joint system, the lateral force developed by the mortar displacement can be of such magnitude to split the brick. The definition of this lateral stress was proposed as [9]:

$$
\Delta S_{xb} = \frac{\Delta S_y \left[\upsilon_b - \dfrac{E_b}{E_m\,(S_1,S_3)}\,\upsilon_m\,(S_1,S_3) \right]}{\left[1 + \dfrac{E_b}{E_m\,(S_1,S_3)}\,\dfrac{t_b}{t_m} - \upsilon_b - \dfrac{E_b}{E_m\,(S_1,S_3)}\,\dfrac{t_b}{t_m}\,\upsilon_m\,(S_1,S_3) \right]}
\tag{8.9}
$$

where ΔS_{xb} = increment of lateral tensile stress in the brick; ΔS_y = increment of vertical compressive stress on the brick; υ_b = Poisson's ratio of the brick; E_b = Young's modulus of the brick; S_1 and S_3 are the vertical and lateral principal stresses, respectively, in the mortar; $\upsilon_m\,(S_1,S_3)$ = Poisson's ratio of the mortar as a function of the mortar principal stresses; $E_m(S_1,S_3)$ = Young's modulus of the mortar as a function of the mortar principal stresses; t_b = thickness (height) of brick; and t_m = thickness of mortar bed joint. Equation 8.9 describes the increment of lateral stress in the brick, resulting from an increment of compressive stress applied to the brick. The lateral stress in the brick is a function of the material properties of the brick and mortar. Poisson's ratio, υ_m, and Young's modulus, E_m, of the mortar are expressed as a function of the vertical stress, S_1, and the lateral stress, S_3. This accounts for the non-linear properties of the mortar with respect to the existing state of stress. The properties of the brick are assumed to be constant under all stress states. Figure 8.10 defines the X-Y coordinate system of the brick used in Equation 8.9. The mortar stress S_1 is in the Y direction while the S_3 mortar stress is basically an axis-symmetric radial stress in the horizontal plane of the X axis.

Equation 8.9 assumes the lateral stress is uniform through the brick thickness t_b. The lateral stress in the brick is a result of the accumulation of the lateral shear stress between the brick and mortar interface. Interface shear stress would be zero at the centerline of the brick and mortar joint. The interface shear stress increases out to the ends of the joint. As a result, the maximum uniform lateral tensile stress (S_x) in the brick will be at the brick centerline.

McNary and Abrams [9] also provide interesting compressive stress-strain data on mortar. Figure 8.15 describes their tests on a masonry mortar defined as Type M. Here compressive stress-strain tests were conducted as with the Sample B refractory mortar tests. However, with the masonry mortar tests, a compressive confining stress, S_3, was also applied to the mortar specimens. Note that increasing the compressive confining stress, S_3, from 0.2 MPa to 6.90 MPa increased the maximum axial compressive stress from about 30 MPa up to about 70 MPa. Also, the maximum strain correspondingly increased from about 0.5% to about 1.40%. These

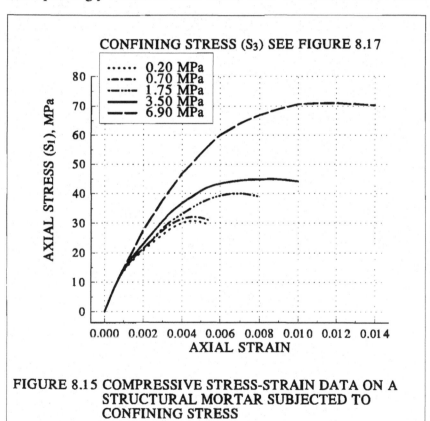

FIGURE 8.15 COMPRESSIVE STRESS-STRAIN DATA ON A STRUCTURAL MORTAR SUBJECTED TO CONFINING STRESS

data confirm that the refractory mortar joint confining stress increases the strength of the refractory mortar. The confining stress in the refractory mortar joint accounts for the difference in the mortar behavior obtained from the split refractory mortar joint specimen tests and the mortar specimen tests for the refractory Sample B mortar.

Figure 8.16 describes the result of the strain measurements for the mortar specimen tests as described in Figure 8.15. The lateral and axial strain curves are used to evaluate Poisson's ratio. For a linear curve, Poisson's ratio (υ) for a given confining stress curve is defined as:

$$\upsilon = \varepsilon_L / \varepsilon_A \qquad\qquad (8.10)$$

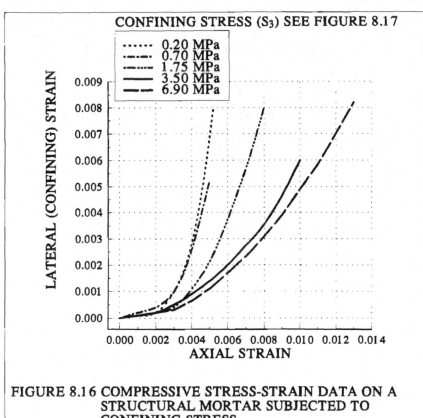

FIGURE 8.16 COMPRESSIVE STRESS-STRAIN DATA ON A
STRUCTURAL MORTAR SUBJECTED TO
CONFINING STRESS

where ϵ_L and ϵ_A are the lateral and axial strains for any given point on the curve in question. For non-linear curves, the instantaneous slope is used.

With the mortar data in Figures 8.15 and 8.16, the stress-dependent MOE can now be calculated. These calculations would be similar to the calculations described for the Sample A and B refractory mortar. The basic mortar joint stresses (axial S_1 and lateral S_3) are illustrated in Figure 8.17.

Figure 8.18 describes the results and shows how the mortar MOE varies as a function of the axial and lateral stresses. This information provides additional insight into the results obtained from the refractory mortar tests and adds credence to the idea of a stress-dependent MOE for refractory mortar.

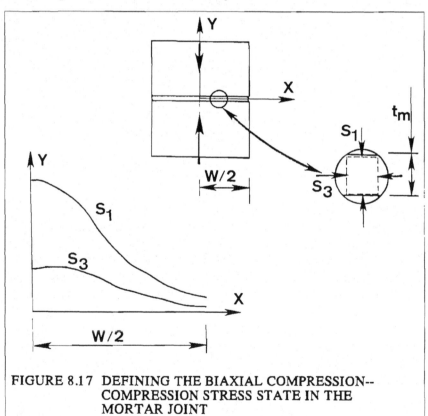

FIGURE 8.17 DEFINING THE BIAXIAL COMPRESSION--
COMPRESSION STRESS STATE IN THE
MORTAR JOINT

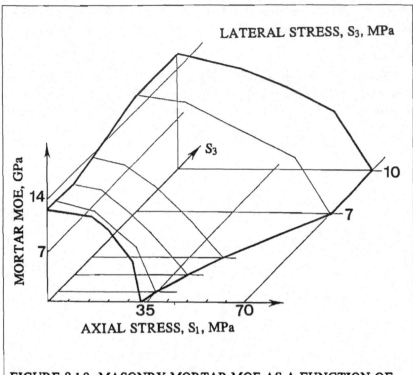

LATERAL STRESS, S₃, MPa

FIGURE 8.18 MASONRY MORTAR MOE AS A FUNCTION OF
LATERAL AND AXIAL STRESS

McNary and Abrams made one additional observation regarding the relationship of the masonry brick ultimate strength with respect to the biaxial tension-compression stress states. They cite the work by Khoo [10] which describes how the tensile stress, developed in the brick by the mortar joint, can greatly affect the ultimate compressive crushing strength of the brick. Khoo non-dimensionalized the data with respect to the uniaxial compressive strength, f_c', and the direct tensile strength, f_t'. The data fall in a narrow, concave band for each brick type and test method. The line of best fit (least squares) through all of the data can be expressed using the equation:

$$S_c / f_c' = 1 - S_t / f_t' \qquad\qquad (8.11)$$

where S_t = the tensile stress in brick and S_c = the compressive stress in brick. The failure curve is described in Figure 8.19 for several brick prisms subjected to a range of biaxial tensile and compressive stress states. This failure information provides additional insight into possible modes of failure in refractory brick subjected to biaxial tension-compression stress states.

B. Summary

The results of the structural masonry mortar joint study provide valuable insight into the thermomechanical behavior of refractory mortar joint systems. The investigative work on these masonry joint provides additional evidence with regard to refractory mortar joint behavior. The following conclusions can be made:

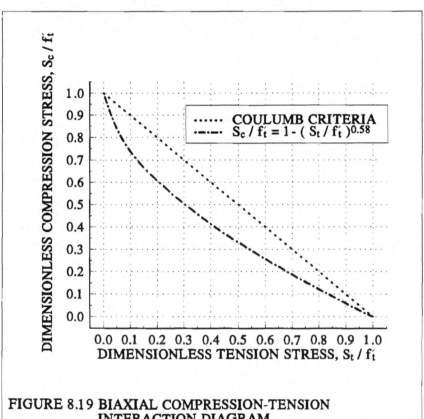

FIGURE 8.19 BIAXIAL COMPRESSION-TENSION
INTERACTION DIAGRAM

a.   The mortar confinement is an important factor with respect to mortar strength.

b.   The mortar specimen compressive stress-strain tests without the use of confining pressure or boundaries do not appropriately evaluate the true mechanical behavior of the mortar in the mortar joint.

c.   The confinement of mortar increases both the compressive strength and the stiffness (MOE) of the mortar.

d.   Perhaps the most appropriate method of evaluating the mechanical behavior of mortar is by testing mortar in an actual mortar joint using split mortar joint specimens.

VI. INFLUENCE OF MORTAR JOINT THICKNESS ON MORTAR JOINT BEHAVIOR

There is very limited literature with regard to the influence of the mortar joint thickness t_m on the mechanical behavior of the mortar joint. The available data [2] suggest that as the mortar joint thickness increases, the MOE of the composite system (brick plus mortar joint) does not continue to decrease or to soften. The studies conducted were suppressed-expansion experiments. Figure 8.20 is taken from Ref. 11, which summarizes the results of the composite MOE determined from mortar joint thicknesses ranging from 1 to 4 mm. The composite MOE is for the test temperature of 450°C. The tests also used a variety of mortar types. Mortar Types A through D were heat-set mortars. Mortar Types E through H were air-set mortars. Mortar Types A through D ranged in grain size, with Type A being the coarsest grade and Type D being the finest grade.   Likewise, in the air-set mortars, Type E was the coarsest grade and Type H was the finest grade. Based on the previous tests discussed in this report, the non-linear behavior of mortar joints may greatly influence the effects of thickening the mortar joint. That is, doubling a mortar joint thickness does not have a corre-

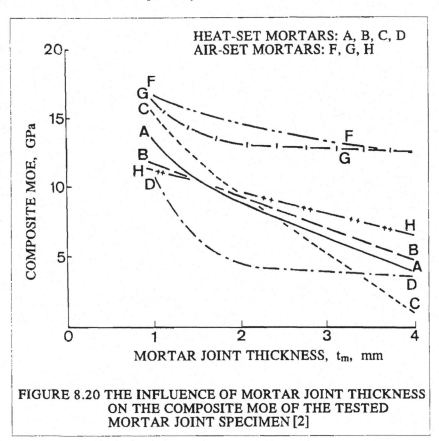

FIGURE 8.20 THE INFLUENCE OF MORTAR JOINT THICKNESS
ON THE COMPOSITE MOE OF THE TESTED
MORTAR JOINT SPECIMEN [2]

sponding linear decrease in the composite MOE.

A comprehensive study was conducted by Koyama, et al. [11] on blast furnace linings. A considerable amount of effort was placed in both elaborate analytical models and an elaborate cylindrical refractory lined vessel test model. The results of the study with regard to mortar joint thickness as related to shell expansion stresses are shown in Figure 8.21. These results agree quite favorably with the BCRA studies summarized in Figure 8.20. Koyama's results also confirm that an increasing mortar joint thickness does not represent a proportional increase in mortar joint expansion allowance.

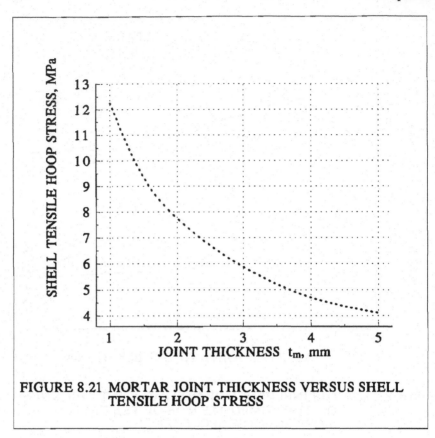

FIGURE 8.21 MORTAR JOINT THICKNESS VERSUS SHELL
TENSILE HOOP STRESS

VII. THE MECHANICAL BEHAVIOR OF
THE DRY JOINT

Compressive stress-strain data on dry (no mortar) refractory joints are extremely rare. Figure 8.22 shows a comparison of compressive stress-strain data on a whole and dry joint sample of a superduty fireclay brick. The data are limited to only two temperatures 480 and 980 °C, however, the *effective* MOE on the dry joint will be evaluated similarly to the procedure used for mortar joints. The dry joint is in contact between the two manufactured surfaces. Most likely, the two dry joint surfaces are not completely flat planes. Therefore, the contact points are experiencing greater compressive stress than those at other locations

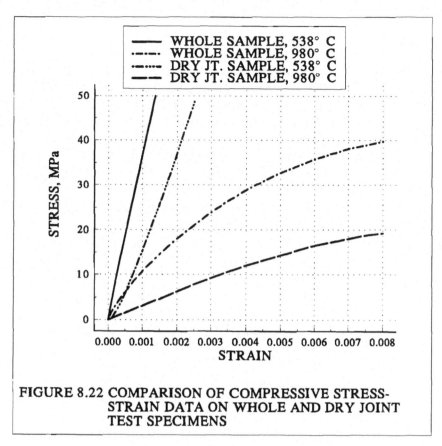

FIGURE 8.22 COMPARISON OF COMPRESSIVE STRESS-
STRAIN DATA ON WHOLE AND DRY JOINT
TEST SPECIMENS

in the specimens. As a result, greater deformations occur at these
contact points.

Comparison of the stress-strain data in Figure 8.22 shows that
the dry joint specimen develops about half the stress as the whole
specimens for any given value of strain. At first look, it appears
that the mortar joint provides more stiffness to the complete lining
of the dry joint.

The effective MOE of the dry joint is defined as E_{dj}. In order
to evaluate the E_{dj} (using Equation 8.5), the thickness of the dry
joint has an assumed value of 1 mm. This is a starting point in
evaluating the behavior of the dry joint.

The calculated dry joint MOE is shown in Figure 8.23. A unique feature of the dry joint is the contact area. As the load increases, the contact area increases. At zero load, no contact area exists. Therefore, the MOE is zero at zero loading. The MOE at 538°C continues to increase with increasing load. This differs somewhat from the lower temperature Sample B mortar MOE (see Figure 8.7). The lower temperature mortar tends to reduce as the stress increases. The Sample A mortar MOE (see Figure 8.3), however, tends to compare favorably with the dry joint MOE. The dry joint MOE at 980°C, however, tends to be significantly less than both mortars' MOE at similar temperatures (see Figure 8.3, 1100°C and Figure 8.7, 800°C). This implies that the dry joint is considerably more flexible at higher temperatures than a mortar joint.

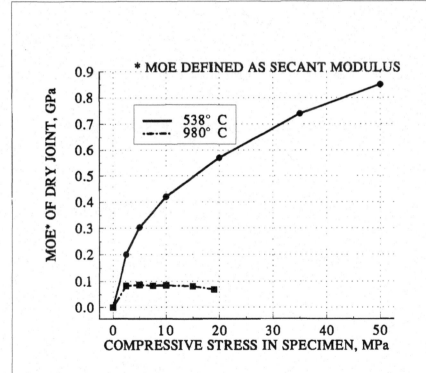

FIGURE 8.23 COMPRESSIVE MOE DATA ON A DRY JOINT
USING A SUPERDUTY FIRECLAY BRICK

It can be concluded that at lower temperatures the small contact area of the dry joint has significant stiffness compared to the mortar joint. However, at higher temperatures, the contact area quickly loses its stiffness compared to the mortar stiffness.

It bears repeating that the dry joint calculated MOE was based on an assumed clearance of 1 mm, with the high point of the surfaces in contact without a compressive loading. The calculated dry joint MOE will vary based on the assumption used for this clearance thickness.

VIII. SUMMARY

The mortar joint is a highly complicated component of the refractory lining system. The use of the mortar joint as an expansion allowance mechanism is appropriate and a necessary part of designing the refractory brick lining system. Some refractory types develop greater expansion forces and, therefore, understanding the thermomechanical behavior of the mortar joint plays a more important role for these type refractories. One should consider whether to include other devices in the design of the refractory lining system, beyond the influence of the mortar joints, to reduce expansion forces. In some cases, the expansion allowance effects of poorly installed lining systems may be too great, reflecting the need for a *tight* lining installation with minimal mortar joint thickness. Thin mortar joints, based on very limited studies, appear to stiffen the lining system at higher temperatures, compared to dry joints.

The behavior of the mortar must be evaluated through laboratory compressive stress-strain tests. Preferably, the mortar should be tested in the mortar joint, as opposed to specimens of the mortar material alone. The confinement of the mortar within the mortar joint is an important factor in evaluating the true mechanical behavior of the system. It is also preferable to have the mortar joint at the joint thickness to be used in the actual lining system.

It can be concluded that the mortar cannot be evaluated in laboratory compressive stress-strain tests when only the mortar has been incorporated into the mortar joint. Compressive stress-strain testing of only mortar specimens has no value with regard to the mechanical design of the lining system.

If there is a need to take advantage of the expansion allowance capability of the mortar joint, then it may be desirable to conduct compressive stress-strain tests on a range of mortar joint thicknesses.

In some industrial environments, the use of thicker mortar joints may be highly undesirable. That is, the use of thicker mortar joints increases the likelihood of the possible penetration of process materials into the joints, resulting in the deterioration of the lining. Minimum mortar joint thicknesses should be used in some cases. Expansion allowance can be achieved by other means.

REFERENCES

1. Hiragushi, K. et al., Property Measurements of Brickwork, Nippon Steel Technical Report, Overseas No. 7, pp. 103-114, November, 1975.

2. Dinsdale, A., Eighteenth Report on the Work of the Refractories Division 1977/78, British Ceramic Research Assoc., September, 1978.

3. James, D. W. F., Nineteenth Report on the Work of the Refractories Division 1978/79, British Ceramic Research Assoc., October, 1979.

4. James, D. W. F., Twenty-First Report on the Work of the Refractories and Industrial Ceramics Division 1980/81, British Ceramic Research Assoc., October, 1981.

5. Schacht, C. A., The Effect of Mortar Joints on the Thermomechanical Behavior of Refractory Brick Lining Systems in Cylindrical Vessels, AISE Spring Conference, Birmingham, AL, March, 1985.

6. Schacht, C. A., An Analytical Look at the Thermomechanical Behavior of Mortar Joints, The American Ceramic Society 95th Annual Meeting, Cincinnati, OH, April, 1993.

7. Tests on Ladle Refractories, BHP Refractories Limited, Report BHP230, 1992.

8. McNary, W. S. and Abrams, D. P., Mechanics of Masonry in Compression, Journal of Structural Engineering, ASCE, Vol. III, No. 4 (1985).

9. Atkinson, R. H., and Noland, J. L., A Proposed Failure Theory for Brick Masonry in Compression, Proceedings, 3rd Canadian Masonry Symposium, Edmonton, Canada, 1983, pp. 5-1 to 5-17.

10. Khoo, C. L., A Failure Criterion for Brickwork in Axial Compression, Ph.D. thesis, University of Edinburgh, Edinburgh, Scotland, 1972.

11. Koyama, Y., Iryama, M., Uchiyama, S. and Imabeppu, M., Investigation of Wear Mechanism of Lining at Lower Stack of Blast Furnace and Measure of Its Prevention, Nippon Technical Report, Overseas, No. 35, pp. 1-13, 1982.

9
Refractory Lining Hinges

I. INTRODUCTION

The refractory lining systems that are defined as arches, domes or other refractory shell-type structures develop, upon heat-up, *hinges*, or pivot points, within the lining. For the lining made up of refractory shapes, the hinge results from the separation of a large portion of the joint length due to a tensile displacement condition that is transferred across a large portion of the joint. The result being that one portion of the joint is in compression while another portion is separated. A more detailed discussion follows with regard to the force and moment details of hinges.

The hinge locations within a refractory shell-type structure are not arbitrary. Rather, the hinge locations are governed by the principles of minimum strain energy.

Typically, during lining installation, attention is given to lining joints, keeping them tight and uniform throughout. However, during heat up the refractory lining expands and undergoes a considerable dimensional change. The result is that the dimensionality of the lining system is no longer compatible with the external support structure and the thermally altered dimensional details of the refractory shapes. As a result, in the heated condition, the thermal restraint forces cause a complete redistribution of the stress-strain behavior throughout the refractory structure. The thermally induced dimensional changes also occur in monolithic type refractory lining systems. Because of the weak tensile strength of refractories in general, the tensile fracturing generates joints to allow for these dimensional changes.

The objective of this chapter is to provide the technical basis for the occurrence of joint separation and hinge formation within refractory lining systems. Since higher stress-strain states exist in the regions of the lining that have the greatest joint separation and hinge formation, the strongest refractories should be placed in these regions of the lining system. It is important to know where these high stressed regions are located to design a better refractory lining structure.

II. FUNDAMENTALS OF JOINT BEHAVIOR

The refractory joint exists in all refractory linings made up of refractory shapes. Monolithic linings may have construction joints if the lining is installed in parts. Basically, the construction joints behave in a similar manner to the shape joints. The tensile fracturing of the monolithic lining results in a joint as well. Often fracture initiation grooves can be placed in strategic locations to cause these fracture joints to develop at optimum locations of the monolithic lining.

The fundamental joint is illustrated in Figure 9.1. The joint shown will be used to describe the most basic compressive-tensile displacement behavior of a joint. This joint will be used to represent either a mortared or dry joint. This joint could be from a cylindrical, spherical or any other refractory lining geometry.

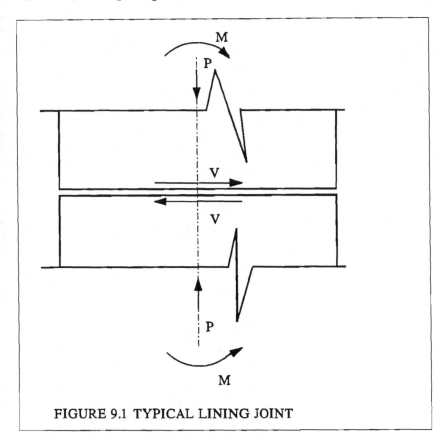

FIGURE 9.1 TYPICAL LINING JOINT

The primary loading across the joint can be defined as a normal load P, a moment load M and a shear load V. Our concern here is with the interaction of the normal load and moment load. The shear load will be addressed separately.

The most fundamental assumption regarding the joint behavior is that the joint is a non-tension interface. Therefore, for a refractory lining made up of refractory shapes, the lining also consists of non-tension interfaces. For most structures that consist of materials such as structural steel or reinforced concrete, the stress-strain states are stresswise linear as described in Figure 9.2. That is, the material can accept either a tensile or compressive stress-strain condition, of equal magnitude. This characteristic of

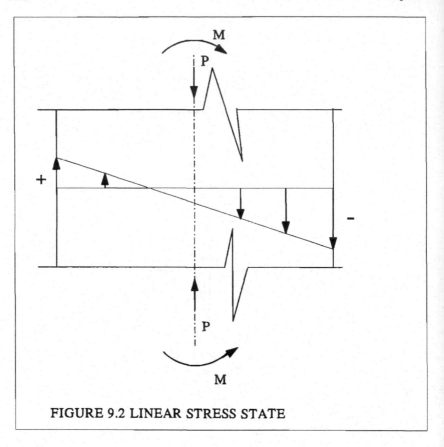

FIGURE 9.2 LINEAR STRESS STATE

the material, having the capacity to develop either a tensile or compressive stress-strain condition, results in a linear geometric behavior of the structure. The geometry of the load-carrying portion of the structure is known prior to the analysis of the structure. It is not required to know the tensile or compressive stress-strain states throughout the structure in order to identify the load carrying portion of the structural geometry. The refractory structure consisting of refractory shapes is not stresswise linear. Since the joint cannot accept a tensile displacement, that portion of the joint which is subjected to tensile displacement does not make up a portion of the load-carrying part of the structural geometry. The following example demonstrates this non-linear aspect of the refractory lining system.

The non-tension condition of a joint is illustrated using the fundamental joint geometry and joint loading shown in Figure 9.1. The joint load consists of a normal compressive load P and a moment load M. These loadings can be redefined by the following equation:

$$e = M/P \qquad (9.1)$$

where e is the eccentricity of the loading P, as measured from the midpoint, center or neutral axis of the joint. The two loads P and M can be redefined and replaced as a single load P at a position at distance e from the joint midpoint (see Figures 9.3a and 9.3b). Both load definitions are identical and the internal equilibrating stress state can be defined as shown in Figure 9.4. The joint

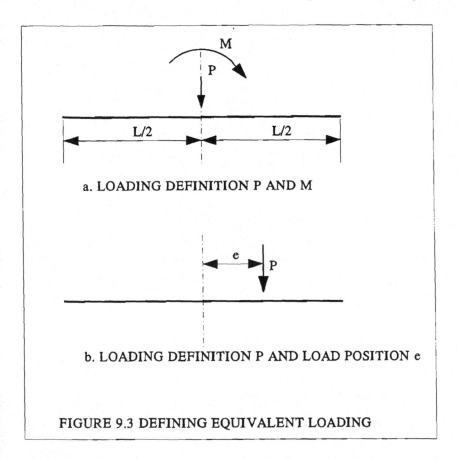

a. LOADING DEFINITION P AND M

b. LOADING DEFINITION P AND LOAD POSITION e

FIGURE 9.3 DEFINING EQUIVALENT LOADING

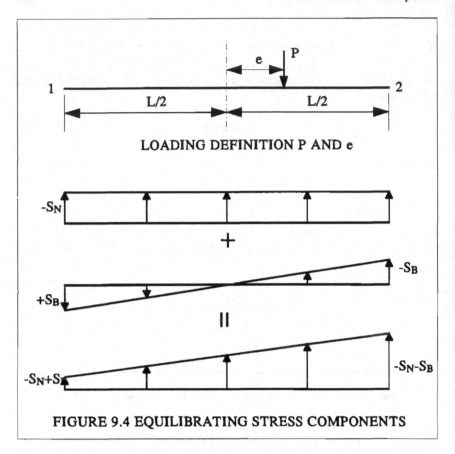

LOADING DEFINITION P AND e

FIGURE 9.4 EQUILIBRATING STRESS COMPONENTS

properties are defined as the joint area A and the joint section modulus S. Assuring a unit depth, the joint area is:

$$A = (L)(1) = L \qquad (9.2)$$

The section modulus is:

$$S = (1)L^2/6 = L^2/6 \qquad (9.3)$$

The normal stress (S_N) is:

$$S_N = P/L \qquad (9.4)$$

The bending stress (S_B) is:

$$S_B = M/S = 6M/L^2 \qquad (9.5)$$

Substituting the value for M from Equation 9.1 into Equation 9.5:

$$S_B = 6Pe/L^2 \qquad (9.6)$$

In Figure 9.4, the stresses at Point 1 (S_1) and Point 2 (S_2) are:

$$S_1 = -P/L + 6Pe/L2 \qquad (9.7)$$

$$S_2 = -P/L - 6Pe/L2 \qquad (9.8)$$

The kern limit is the unique eccentricity e such that the stress at Point 1 is zero. Or (from Equation 9.7):

$$S_A = 0 = P/L - 6Pe/L^2 \qquad (9.9)$$

Solving for e:

$$e = L/6 \qquad (9.10)$$

Equation 9.10 defines the kern limit for a rectangular beam section. The conclusion is that if the eccentricity of the load P is kept within the middle third (L/6) of the rectangular beam cross section, the stress will remain between zero and compressive in value. The stress will not be tensile if the load P is kept within the kern limit.

If the load P is positioned at the kern limit, the maximum compressive stress, S_B, is (from Equation 9.8):

$$S_B = P/L + 6P(L/6)/L^2$$

$$S_B = 2P/L \qquad (9.11)$$

The maximum compressive stress at Point B is twice that of the normal stress when the load is positioned at the kern limit of a beam with a rectangular cross section.

Let us now examine the beam stress state when the load is located outside the middle third of a rectangular beam. In this case, the center of gravity of the beam's internal equilibrating stress is always positioned at the location of the load P. As defined in Figure 9.5, when the load center of gravity is outside the kern limit, a portionof the joint opens working with Figure 9.5. The portion of the joint which is in compression is 3(L/2 - 3). Again, assuming a rectangular cross section of unit width, the maximum compressive stress is evaluated by setting the area of the stress diagram equal to the load P, or:

$$S_B(1/2)[3(L/2 - e)] = P \qquad (9.12)$$

$$S_B = 2P/[3(L/2 - e)] \qquad (9.13)$$

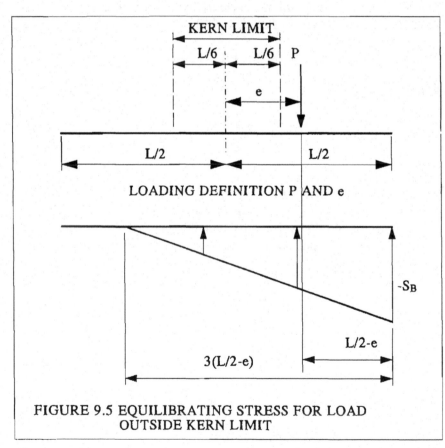

FIGURE 9.5 EQUILIBRATING STRESS FOR LOAD
OUTSIDE KERN LIMIT

If we assume the center of gravity of the loading is at L/4 (e = L/4), then:

$$S_B = 2P/[3(L/2 - L/4)]$$

$$S_B = 8P/3L = 2.5P/L \qquad (9.14)$$

Comparing the maximum compressive stress at Point B, when the load center of gravity shifts from L/6 to L/4, S_B increases from 2P/L to 2.5P/L. As the load eccentricity shifts farther to the right edge, the maximum compressive stress begins to increase rapidly due to the reduction of the bearing area and the reduction of the section modulus.

In summary, when the center of gravity (or eccentricity) of the compressive load remains within the kern limit, the stress is compressive across the full joint length. When the load eccentricity moves outside the kern limit, a portion of the joint opens while the portion in compression correspondingly decreases. As the eccentricity increases, a larger portion of the joint opens with the compressively loaded portion becoming smaller.

The original assumed loading was a normal compressive load and a moment load on the joint. There are corresponding displacements that accompany these two forces. The compressive normal load will result in a compressive displacement of the joint. The moment load will cause a rotation to occur within the joint. As the moment load increases relative to the compressive normal load, causing the eccentricity of the normal compressive load to move outside the kern limit, the compressively loaded portion of the joint decreases with a simultaneous increase in the rotation of the joint. At certain locations within the refractory lining structure, the joints will undergo larger increases in joint opening and rotation. These locations of joint behavior can be interpreted as a *hinge* within the refractory lining.

The assumption of a rectangular beam section is representative of most refractory lining systems. Even for cylindrical or spherical lining systems, the unit section through the thickness is nearly rectangular. The assumption of a rectangular section may not be valid for shell lining with a lining radius-to-thickness ratio of less than 10. As the radius-to-thickness ratio decreases (thick lining relative to radius), the neutral axis of bending is no longer at the mid-thickness, and the through thickness stress variation is no longer linear. In other words, thick-shell behavior begins to govern. Most refractory linings have a radius-to-thickness ratio of 10 or greater. Therefore, thin shell behavior is applicable, as was assumed in the preceding derivations.

Table I [1] provides a review of the kern limit for beam sections of various geometries.

III. HINGE CONCEPT IN REFRACTORY LININGS

The previous discussion dealt with the lining joint and described the progression of the joint from a full length compressively loaded joint to a partial length compressively loaded joint. As the partial length of the compressed portion of the joint decreases, the opposing rotation of the two sides of the joint increases until a pivotal point is achieved. When the joint reaches this degree of rotation the joint can be defined as a *hinge.*

Presented here is an example of a refractory lining structure

Table I

Summary of Kern Limits
for Various Cross Section Geometries

Description of Cross Section Geometry	Definition of Dimension L	Kern* Limit
Rectangular Width b, Depth L	Depth L	L/6
Solid Circular	Radius L	L/4
Thin Wall Circular Shell	Radius L	L/2

*As measured from midpoint of cross section.

with hinge development. The particular refractory lining structure used here is the arch. The arch will be displaced in a symmetric manner which causes the hinge development to occur at the obvious locations. For nonsymmetric loadings, nonsymmetric refractory lining geometry or loadings that are more complicated, the hinge locations are not always obvious. For these later complications the hinge location can only be determined by a structural analysis procedure. Perhaps one of the most common errors by novice analysts of refractory lining structures is with regard to predicting hinge locations. The hinges are arbitrarily assumed to occur at certain locations within the refractory structure without the use of structural analysis. The more reliable refractory materials are then zoned around the assumed hinge locations. Structural analysis may show that hinges occur remote from the assumed locations. For refractory lining structures that are not simplified geometries or that are exposed to complicated load environments, a refractory lining structural analysis may be a prudent choice to evaluate the lining behavior.

The refractory arch shown in Figure 9.6a is used to illustrate the concept of the hinge. As shown, the arch is assumed to be made up of refractory block, or shapes. In this example, the shapes are assumed to be ideally manufactured such that all joints are in full contact upon completion of installation. The arch is symmetric in geometry. The two skews (refractory shapes at each end point of arch) support the arch and are at their initially installed position as described in Figure 9.6a. The loading is assumed to be a vertical downward gravity weight force that is also symmetric.

The arch is assumed to be spread outward at each skew by a displacement of $+\Delta S$, illustrated in Figure 9.6b. The arch in the spread position develops three hinges, one at each skew and one at the arch crown. At the two skews, the hinge point is at the arch interior side while the hinge point at the crown is at the arch exterior side.

The second arch spread condition is an inward displacement at each skew defined as $-\Delta S$ (see Figure 9.6c). In this case, the hinges are at the same location as the previous case except at the opposite sides of the arch.

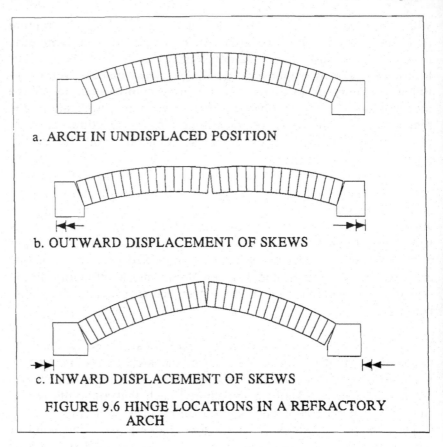

a. ARCH IN UNDISPLACED POSITION

b. OUTWARD DISPLACEMENT OF SKEWS

c. INWARD DISPLACEMENT OF SKEWS

FIGURE 9.6 HINGE LOCATIONS IN A REFRACTORY
ARCH

The arch example provides the obvious solution to hinge
locations within the arch when subjected to a symmetric
displacement condition.   Actually, the last case (equal inward
displacement at each skew) is similar to the case of a heated arch.
The arch is assumed to be uniformly heated from the installed
condition and the skews remain fixed at the installed positions. The
arch will increase in length and as a result will be too long to
accommodate the skew locations. The result being that hinges will
occur as shown in Figure 9.6c. An interesting feature of the arch is
that a statistically determinant condition exists in which the hinge
forces can be evaluated by the simple rules of statics. However, the
hinge locations must be known to evaluate the forces by statics.

The arch example also identifies a rather interesting feature of the refractory lining system. If the arch had been constructed of a homogeneous structural material such as structural steel (no joints), then the displaced skews would not have altered or revised the load bearing portion of the structural. With the structural steel, moments would have been developed at the skew and crown locations rather than hinges. The reason is that structural steel can assist both a tensile or compressive stress-strain environment. The refractory material (with refractory shapes) and the joints cannot resist any significant tensile stress or strain environment. The refractory lining system's load bearing portion of the structure changes as the load environment changes. Initially, the installed arch was supporting the gravity weight loading across the full bearing area of the arch. When the skew displacements are applied, the arch changes from a statically indeterminate to a statically determinant structure. The formation of hinges causes this transition. The hinge formations also change the load bearing area in the region of the hinges.

IV. ADDITIONAL ASPECTS OF HINGE BEHAVIOR

The previous discussions described how the normal compressive load, in combination with the moment load, would create a hinge. The arch example assisted in showing how hinges could be developed within the refractory lining system by imposing a displacement condition and how the pivot point of the hinge changes locations. The assumption was made that the hinge forms one particular joint within the lining system. Actually, the refractory lining joints adjacent to the hinge will have a significant portions that are separated. The following discussion adds additional insight into the joint behavior in the region of the hinge.

As previously discussed, the portion of the joint that is exposed to a compressive loading is a function of the eccentricity of the loading, as expressed in Equation 9.1. The peak eccentricity will result in the location of the hinge. As illustrated in Figure 9.7 and 9.76, the normal compressive load and the moment load will most likely vary along the length of the refractory structure, as described in Figure 9.7c. The dithered area in Figure 9.7b represents

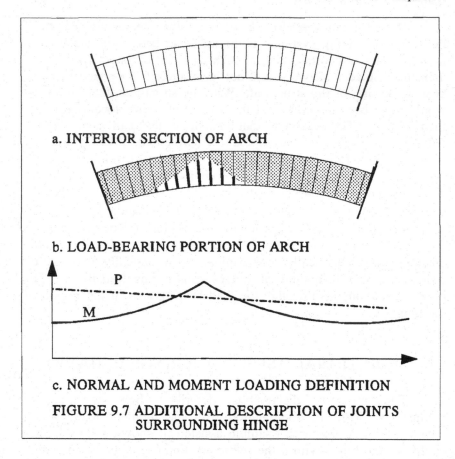

a. INTERIOR SECTION OF ARCH

b. LOAD-BEARING PORTION OF ARCH

c. NORMAL AND MOMENT LOADING DEFINITION

FIGURE 9.7 ADDITIONAL DESCRIPTION OF JOINTS
SURROUNDING HINGE

the joints that are in full-length compression. That is the load P is
within the kern limit. In the region of the hinge, the load P is
outside the kern limit. At the point of the maximum moment M,
the eccentricity of the load P is a maximum. The heavy-lined joints
represent the separated portion of the joint. The eccentricity will
decrease from each side of the hinge due to the decrease in the
moment load, as described in Figure 9.7c. The portion of each
lining joint that is subjected to the compressive loading defines the
portion of the lining that represents the load-carrying part of the
lining.

The portion of the lining that is defined by the separated
joints does not contribute to supporting the lining, but simply adds

the gravity loading to this portion. This portion of the lining also provides the necessary thermal heat containment.

The lining structure has regions that provide both lining support and heat containment functions, while other parts of the lining provide only a heat containment function. It is, therefore, necessary in a refractory lining structural investigation to identify those parts of the lining that support the lining system such that the refractory materials with the highest degree of structural integrity are placed in these regions.

V. AN ANALYTICAL LOOK AT HINGE FORMATION

The previous discussions on hinge formation addressed the nature of specific forces necessary to cause a hinge to occur as well as the general nature of the joint behavior for those joints near the hinge location. The following discussion introduces the strain energy concept used to locate hinges within a lining system. The previous arch example was used to demonstrate the concept of hinge formation. The following example is applicable to lining systems which are not geometrically symmetric or that are exposed to non-symmetric loadings. In these cases, the location of the hinges are not so obvious.

The strain energy concept is a fundamental part of structural analysis theories [2-5] and is a basis for many structural analysis procedures. The finite element method is basically a strain energy method. The finite element analysis method is used to duplicate the results in the following example.

Though the stone arch had been in usage for centuries, British engineers conducted experimental investigations on the stone arch in the early 1900s. Their interest in arch behavior was due to the usage of the stone arch as a vital highway bridge throughout Great Britain. The foundations at each end of the stone highway arch tend to settle and displace over time. As a result, concerns were raised over the hinges that were formed within the arch due to the support displacement.

Two types of loads are imposed on the highway bridge arch. The first loading is the symmetric vertical gravity load of the arch mass and the highway material's overburden. This load would cause the hinge to locate at the crown of the arch for the arch with spread foundations. The second loading is a moving vehicular load. This is the loading of concern. As the vehicular loading moves across the highway, the hinge changes location because of the changing vehicular load location. The basic results of their investigation are summarized here. A finite element study was also conducted to determine if the hinges' transient nature, as reported in the British experimental studies, could be duplicated analytically.

Figure 9.8a describes the experimental arch that was constructed of machined steel blocks. Each block is an equal arc

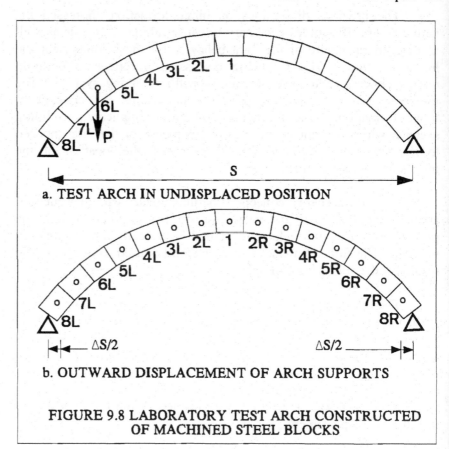

a. TEST ARCH IN UNDISPLACED POSITION

b. OUTWARD DISPLACEMENT OF ARCH SUPPORTS

FIGURE 9.8 LABORATORY TEST ARCH CONSTRUCTED
OF MACHINED STEEL BLOCKS

segment of the arch. The arch consists of a total of 15 blocks. Details of the experimental arch are provided in Reference 6. The arch is 75 mm (3 in.) thick (in radial direction), 38 mm (1 1/2 in.) wide and has a span (S) of 1220 mm (48 in.). Thin sheets of rubber, 1/4 mm (0.01 in.) thick, were placed between each block to assure uniform bearing surfaces. The arch had a radius of 762 mm (30 in.) as measured to the block mid-depth. The blocks are shown numbered sequentially starting from the crown joint, in the left and right directions (see Figure 9.8a). The analytical arch was constructed with pinned ends and made of triangular elements. A support pin was also installed at the center of each block (circle at center of blocks in Figure 9.8a) to facilitate the hanging of a weight, simulating the overburden material loading and the vehicular loading. By changing the vehicular weight location, the simulation of a vehicular load traveling across the road at the top slag of the arch was achieved.

The laboratory tests were conducted by first spreading the base of the arch a small amount outward (ΔS) in order to create a hinge within the arch, as illustrated in Figure 9.8b. With only the vertical gravity load on the arch, the hinge occurred at the arch crown with the pivot point at the outside or top surface of the block. The hinge would form at either side of block number 1. A hinge would not form at each side since a four-hinged arch is unstable. The four-hinged arch would revert to a three-hinged arch. It should also be noted that the small ΔS spreading does not significantly change the arch dimensions. The weight simulating the vehicular weight was then hung at the various block pin positions, and the resulting hinge location was recorded.

The analytical study was limited to the experimental test results for the vehicular load over block 6L. The vehicular concentrated weight was, therefore, hung on the pin of block 6L.

Two analytical investigation cases [8] were conducted for the vehicular loading at block 6L. Case I replicates the experimental test in which the vehicular loading (P) was increased from a very small value, for which the vehicular load was negligible compared to the total overburden material weight, to a high value, for which the vehicular load was very significant compared to the overburden material weight. The purpose of Case I is to observe and confirm the repositioning of the hinge as a function of the vehicular weight.

The Case II investigation concentrated on the summed strain energy within the arch as a function of the vehicular weight and the hinge location. For Case II, the arch was analyzed with the hinge fixed at a location between the blocks while the vehicular load was increased. The analysis was then repeated for the hinge fixed at each of the various locations between two blocks, ranging from the crown of the arch to the adjacent position of the load. The hinge position defined as 1L-2L implies the hinge is located between blocks 1L and 2L. In all cases the hinge was located on the top side of the arch.

The finite element model used to analyze the experimental arch is described in Figure 9.9 [8]. This model was mathematically constructed of constant strain triangular (CST) elements. At the time of this work the CST element was recognized as a reliable finite element and was used to define the arch model. Each arch block is made up of 32 CST elements. For the Case I analysis, nonlinear interface elements were used at each joint, as shown schematically by the short dash in Figure 9.10. A total of five interface elements were used to simulate the joint behavior. The nonlinear characteristics of the interface elements are that only a compressive load can be transferred across the joint. The interface element will not transfer a tensile force across the joint. If a tensile force is developed, the interface will cause the joint to separate and a zero force condition to occur at that location of the joint. Because of this non-linear behavior, it is necessary to iterate or conduct successive solutions, by gradually ramping the loads and displacements until the interface behavior has converged to a repeated behavior. At the start of the analysis all interface elements are assumed to be closed. Through the sequential analysis procedure, the open or closed condition of each interface element is determined.

A. Results of Case I Investigation

The vehicular load P was applied at the center of block 6L. This load was increased from a zero value to a maximum value of

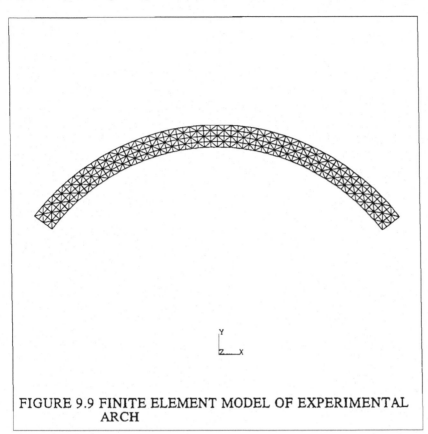

FIGURE 9.9 FINITE ELEMENT MODEL OF EXPERIMENTAL
ARCH

44.10 Newtons. The arch was spread outward at each base by a total amount of 0.008 mm (0.0003 in.). The vehicular loading was sequentially increased over a total of thirty increments. This was necessary to avoid a mathematical collapse of the model.

The results of the analysis showed that at the maximum loading of 42.30 Newtons, the hinge converged to the 5L-6L location. The experimental investigators failed to record the value of the overburden material loading. Therefore, it was not possible to compare the vehicular load value versus hinge location between the experimental and analytical studies. However, the analytical study confirmed the experimentally observed behavior that the hinge position does change as a function of the magnitude of the vehi-

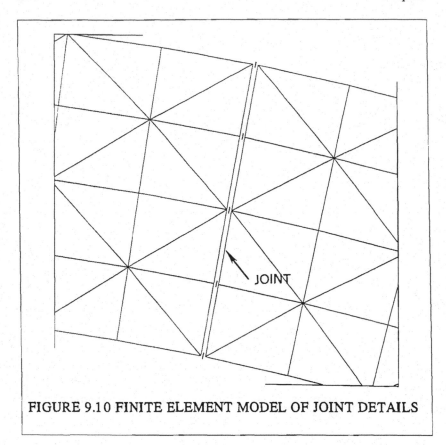

FIGURE 9.10 FINITE ELEMENT MODEL OF JOINT DETAILS

cular load. The results also confirmed the methodology of using the interface elements to duplicate the changing hinge location. The magnitude of the load versus the hinge location is summarized in columns 1 and 2 of Table II.

B. Results of Case II Investigation

For the Case II investigation the model was altered at the joints. The joints were coupled except for the hinge location. Therefore, a three-hinged arch (one hinge at each support) was evaluated. The primary interest in this investigation was the magnitude of the strain energy for an increasing vehicular load. Separate analyses were conducted for the hinge at locations 1L-2L,

Table II

Comparison of Load P Results of Experimental Study [6] and Finite Element Study [8]

Position of Hinge	Value of Load P for Arch Model Study [6] (Newtons)	Value of Load P for Three-Hinged Arch Energy Study [8] (Newtons)
1L-2L	0 to 23.85	0 to 24.75
1L-2L to 2L-3L	23.85	24.75
2L-3L to 3L-4L	36.90	38.25
3L-4L to 4L-5L	42.30	44.10
4L-5L to 5L-6L	42.30	44.10
Horizontal Displacement ΔS at Support = 0.0003 in.		

2L-3L, 3L-4L, 4L-5L and 5L-6L. For each hinge location, the vehicular load was increased sequentially.

The total strain energy was recorded for each value of vehicular load. For each of the analyses the strain energy (SE_i) was evaluated as defined by:

$$SE_i = E / 2 \sum_{n=1}^{M} e_n^2 V_n \qquad (9.15)$$

where the subscript i designates the hinge location at joint i, e_n is the strain within the element n of the model, E is the modulus of elasticity, M is the total number of elements within the analytical model and V_n is the volume of element n. The results are plotted in Figure 9.11.

The strain energy results in Figure 9.11 are plotted for each hinge location. As shown, the two curves for hinges 1L-2L and 2L-3L intersect at a vehicular load value of 24.75 Newtons. As this load increases beyond the 24.75 value, the lesser strain energy

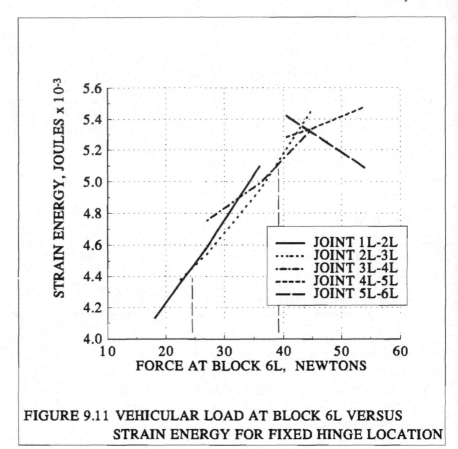

FIGURE 9.11 VEHICULAR LOAD AT BLOCK 6L VERSUS
STRAIN ENERGY FOR FIXED HINGE LOCATION

occurs when the hinge is at 2L-3L. Therefore, if the hinge was
allowed to change, it would shift from 1L-2L to 2L-3L when the
vehicular load increases past the value of 24.75. This rationale
regarding hinge location is observed with the increasing value of
the vehicular load beyond the value of 38.25 and 44.10. The
results of the Case II investigation are summarized in columns 1
and 3 of Table II. The Case II results agree favorably with the
Case I results. Most likely the small differences between the Case I
and II results arise from the partial joint openings that occured in
the Case I model. That is, in Case I the hinge is formed
automatically within the arch. The joints adjacent to each side of
the hinge are partially separated. In Case II the hinge location is
specified with the remainder of the joints totally closed. As a

result, the strain energy will be slightly different between Case I and II for a given hinge location.

VI. SUMMARY

Refractory linings are constructed from shapes of various geometric properties in order to form the desired lining geometry. The resulting joints result in a complete structural behavior. Lining joints, mortared or dry, cannot sustain any significant tensile strain. Lining joints can only sustain compressive loads. This stress-dependent behavior of joints results in the formation of hinges within lining systems. Hinges typically occur in shell type refractory lining systems defined as arches and domes. Because of nonsymmetric lining geometry and/or nonsymmetric loadings, the location of hinges within the lining system are not always obvious, and can be difficult to predict. Modern analytical structural techniques, such as the finite element method, provide a valuable tool in predicting the lining behavior and in locating the hinges.

Monolithic linings are also weak in tension. Hinges will also form in monolithic linings because of the lack of tensile strength. These hinge positions can also be predicted just as with the brick linings.

REFERENCES

1.   Roark, R. J. and Young, W. C., Formulas for Stress and Strain, Fifth Edition, McGraw-Hill Book Company, 1975, pp. 440-441.

2.   Carpenter, S. T., Structural Mechanics, John Wiley & Sons, Inc., 1960.

3.   Timoshenko, S. and Young, D. H., Theory of Structures, McGraw-Hill Book Company, Inc., 1945.

4.   Crandell, S. H., Engineering Analysis: A Survey of Numerical Procedures, McGraw-Hill Book Company, Inc., 1956, pp. 18-26.

5.   Prezemieniecki, J. S., Theory of Matrix Structural Analysis, McGraw-Hill Book Company, 1968.

6.   Pippard, A. J. S., Tranter, E., and Chitty, L., The Mechanics of the Voussoir Arch, Journal of the Institution of Civil Engineers, ICE, London, Vol. 4, Paper No. 5108, December, pp. 281-306 (1936).

7.   Pippard, A. J. S. and Ashby, R. J., An Experimental Study of the Voussoir Arch, Journal of the Institution of Civil Engineers, ICE, London, Vol. 10, Paper No. 5177, January, pp. 383-404 (1936).

8.   Schacht, C. A., A Viscoelastic Analysis of a Refractory Structure, Ph.D. dissertation, Carnegie-Mellon University, 1972.

10
The Influence of Stress State on the Strength of Refractories

I. INTRODUCTION

Another complicating factor in dealing with refractories is the strength. The ultimate crushing strength or the ultimate tensile strength of a refractory will vary with the dimensionality of the stress state, the temperature and the rate the loading is applied to the refractory material. These traits are not unique to refractories, but are common to brittle type materials in general. Perhaps the most researched material with regard to the effects of stress states on ultimate strength is structural concrete. The objective of this chapter is to inform the refractory user regarding the influence of these factors on the ultimate strength and to provide basic guidelines for determining whether a refractory lining will fail based on ultimate uniaxial crushing strength or uniaxial tensile strength data.

II. BACKGROUND

Strength data typically provided by refractory manufacturers to quantify the compressive strength of a refractory material are the ultimate crushing strength and the modulus of rupture (MOR) for the ultimate tensile strength. In testing the former, the compressive load is increased until the specimen fails. The estimated ultimate tensile strength (or MOR) is evaluated by a beam specimen in which the ultimate tensile bending stress is evaluated. Both of the above tests are conducted using ASTM-defined test procedures. Both of these tests evaluate the ultimate strength (tension and compression) of a specimen subjected to a uniaxial (or one-dimensional) stress state. Chapter 4 provides details of the ASTM test procedures.

Our concern in this chapter is using the uniaxial strength data as a guide in determining if the determined refractory lining stress state is satisfactory or unsatisfactory. Typically, the lining stress state, at any point of interest, is a multidimensional stress state. Investigations of brittle material's ultimate strength have shown that a multidimensional stress state has a profound influence on the ultimate strength of brittle materials. That is, a compressive multidimensional stress state can have stresses that exceed the ultimate compressive stress obtained from a uniaxial compressive strength test. A brief summary is provided on the influence of multidimensional stress states with regard to the ultimate strength.

Other factors that influence the ultimate strength of refractories are the temperature of the refractories and the rate of the loading. Both of these influences on the ultimate strength of refractories will be discussed.

Nearly all of the investigative work on the influence of the dimensionality of the stress state on ultimate strength has been conducted on structural concrete. This is expected since concrete structures must be designed for a variety of loading conditions. A design stress must be determined such that a factor of safety against failure is established. Very little investigative work has been done on the influence of the dimensionality of the stress state on the ultimate strength of refractories. Most of the work on the influence of temperature on ultimate strength has been done on refractory materials.

III. TENSILE STRENGTH

The ultimate tensile strengths of refractories are significant for the objective of determining if fracturing and resulting deterioration of the refractory lining will occur. However, as discussed in Chapter 5, it is also important to know the ultimate tensile strain since tensile fracturing is caused primarily by the thermal loading (or strain-controlled loading). However, since testing procedures are only set up for determining the ultimate tensile stress, only it will be discussed here.

A. Multidimensional Stress States

Most investigators [1-10] of brittle materials have shown that the ultimate tensile strength is not influenced by the dimensionality of the stress state. The references on concrete are presented as only a starting point on material strength as influenced by the stress state. Considerable investigative work has been conducted on concrete strength in the past two decades. Biaxial (two-dimensional) or triaxial (three-dimensional) tensile stress states will cause tensile fracturing at tensile stress levels identical to the uniaxial (one-dimensional) ultimate tensile strength. Figures 10.1 and 10.2 describe the ultimate stress (tensile or compressive) envelopes for two- and three-dimensional stress states, respectively, in brittle materials. Note, the ultimate compressive stress in brittle materials is typically much greater than the ultimate tensile strength.

Figure 10.3 provides an amplification of the two-dimensional tensile stress region of Figure 10.1 [6]. As shown in Figure 10.3, either one of the two principal tensile stress components (S_1 and S_2) is no greater than the ultimate uniaxial tensile strength (f'_t) when the adjoining two-dimensional stress component is in the range of f'_t to a compressive stress that is a low ratio of f'_t (3 to 8 as shown for the particular material in Figure 10.3). The failure envelope is identified by the heavy line adjacent to the two principal stress axes S_1 and S_2. The assumption used in this failure envelope was that

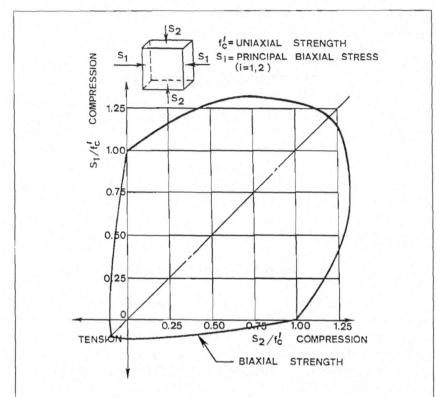

FIGURE 10.1 THE INFLUENCE OF A TWO-DIMENSIONAL
STRESS STATE ON ULTIMATE STRENGTH

the uniaxial compressive strength was 8 times the absolute value of the ultimate tensile strength. This ratio affects the manner in which the tensile failure envelope transitions into the compressive stress portion of the envelope. However, this ratio does not influence the portion of the envelope in the total tensile stress region.

As also shown in Figure 10.3, the transition of either of the stress components from a tensile to a compressive stress results in a decreasing ultimate tensile stress of the opposing tensile stress components. In Figure 10.3, the square geometry of the failure envelope in the tensile-tensile portion of the two-dimensional stress state assumes that the inherent cracks within the specimen are parallel to one of the unstressed axes. If this assumption is

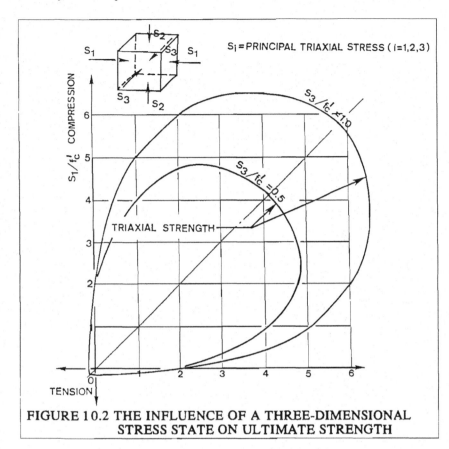

FIGURE 10.2 THE INFLUENCE OF A THREE-DIMENSIONAL
STRESS STATE ON ULTIMATE STRENGTH

removed, then the failure envelope is defined by the broken line. The cross coupling of the two-dimensional tension and compressive stress was briefly addressed in Chapter 8 (see Figure 8.19). This tensile-compressive portion of the failure envelope agrees fundamentally with the failure envelope of Figure 10.3.

It can be concluded that the dimensionality of the stress state does not influence the magnitude of the tensile failure stress. The maximum tensile failure stress is not amplified due to the dimensionality of the stress state. When one of the stress components become compressive, the maximum tensile failure stress is reduced to a level less than the ultimate tensile stress. The introduction of a third stress component does not alter these conclu-

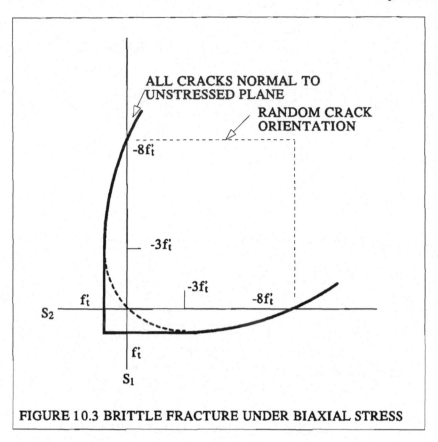

FIGURE 10.3 BRITTLE FRACTURE UNDER BIAXIAL STRESS

sions.   The triaxial stress state tensile failure criterion is similar to
the biaxial tensile failure criterion.

B. Effect of Temperature on MOR

Investigation on the effects of temperature on refractory
tensile strength (hot MOR, or HMOR) shows that some refractories
gain considerable strength at elevated temperatures. The following
discussions present the results of several investigators of refractory
HMOR.

HMOR data [11] on basic refractory brick are described in
Figures 10.4, 10.5 and 10.6. These data show the magnesite chem-

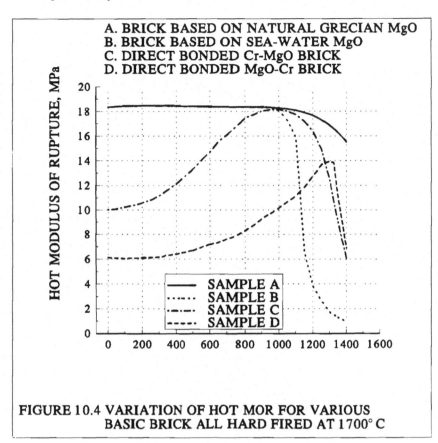

A. BRICK BASED ON NATURAL GRECIAN MgO
B. BRICK BASED ON SEA-WATER MgO
C. DIRECT BONDED Cr-MgO BRICK
D. DIRECT BONDED MgO-Cr BRICK

FIGURE 10.4 VARIATION OF HOT MOR FOR VARIOUS
BASIC BRICK ALL HARD FIRED AT 1700° C

istry, the type of bonding (fired or chemical) and the magnitude of the firing temperature influencing the HMOR. The effects of peak firing temperature on direct bonded chrome-magnesite brick are compared to a chemically bonded chrome-magnesite brick in Figure 10.5. As shown, the less the firing temperature, the less the peak MOR stress. Here, the data curves for the samples fired at 1700 and 1550°C overlap for temperatures below 500°C. The chemical bonded stress has the lowest MOR stress at the higher temperature range, but the highest MOR stress at the lower temperature range. All of the chrome-magnesite brick MOR stresses fall to minimum values at 1400°C. For the peak firing temperatures, the peak MOR stress is in the range of 80 to 1200°C.

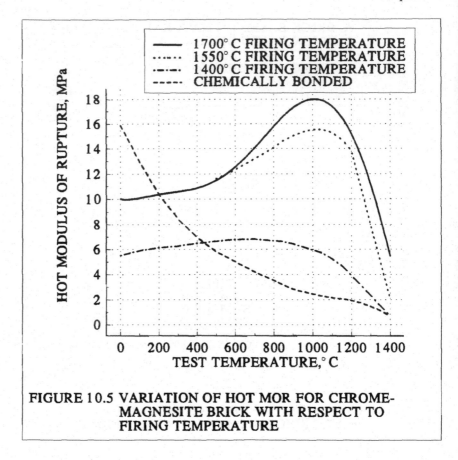

FIGURE 10.5 VARIATION OF HOT MOR FOR CHROME-
MAGNESITE BRICK WITH RESPECT TO
FIRING TEMPERATURE

Figure 10.6 provides a comparison on magnesite-chrome bricks similar to that discussed for the chrome-magnesite bricks. The HMOR stress behavior of the magnesite-chrome brick is somewhat similar to that of chrome-magnesite brick. The firing temperature has a more profound effect on the peak MOR stress. Also, the peak MOR stress for the direct bonded brick is over a narrower temperature range and peaks at a higher temperature. The data curves for the firing temperatures of 1700 and 1550°C overlap for temperatures less than 700°C. The data curves for the 1400°C fired sample and the chemically bonded sample overlap for temperatures greater than 600°C.

The influence of temperature on the MOR stress of fireclay

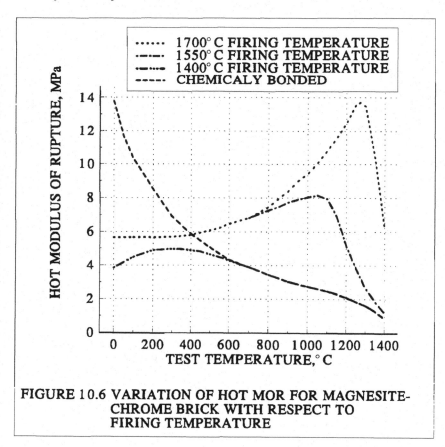

FIGURE 10.6 VARIATION OF HOT MOR FOR MAGNESITE-CHROME BRICK WITH RESPECT TO FIRING TEMPERATURE

and alumina brick [12] are described in Figures 10.7 and 10.8.

In Figure 10.7, the peak MOR stress is reached at about 1100°C. The MOR stress is uniform from room temperature up to about 850 to 950°C. Beyond 1100°C the MOR stress decreases to minimum values at about 1400°C.

Figure 10.8 describes the effects of temperature on the MOR stress for high purity and density high-alumina brick. There is some variability in the results. A portion of the dense high-alumina brick (density 60 and 90% alumina) has a peak HMOR stress at 1100°C, while the 99% high alumina brick HMOR stress continued to decrease with increasing temperature. The superduty fireclay data

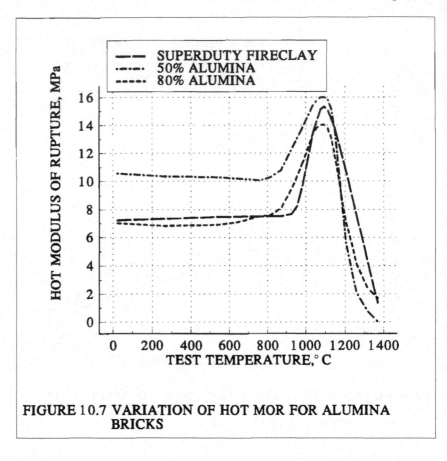

FIGURE 10.7 VARIATION OF HOT MOR FOR ALUMINA
BRICKS

are included in Figure 10.8 as a reference for the magnitudes of the
high-alumina HMOR.

The temperature dependency of MOR stress for silica and
semi-silica brick is described in Figure 10.9. The silica brick,
unlike the alumina brick, is not greatly influenced by temperature.

According to the investigators of the MOR data in Figures
10.7, 10.8 and 10.9, their conclusions were:

1.   The shape of the strength-temperature curve was

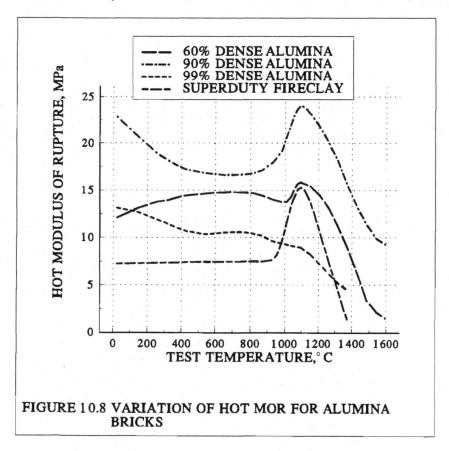

FIGURE 10.8 VARIATION OF HOT MOR FOR ALUMINA
BRICKS

controlled by the presence or absence of disimilar mineral phases possessing different thermal expansions.

2.   The high-temperature modulus of rupture of brick in the alumina-silica system was largely controlled by the purity of the brick and their apparent porosity.

Figure 10.10 [13] describes the temperature-dependent MOR stress of 80-85% alumina brick. All were fired brick except Sample No. 6, which was phosphate bonded. The phosphate bonded brick exhibited high MOR strength at low temperatures, as

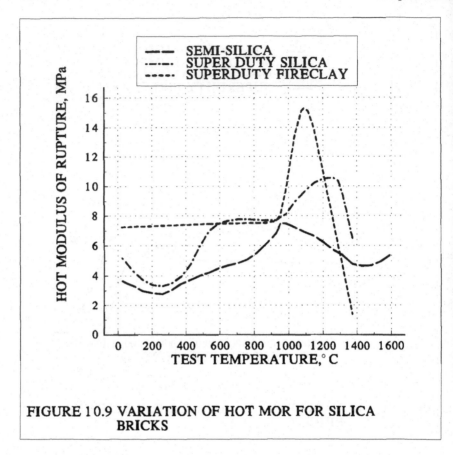

**FIGURE 1 0.9 VARIATION OF HOT MOR FOR SILICA
           BRICKS**

seen by Miller and Davies [12]. The Greaves MOR data [13] in
Figure 10.10 showed the lowest test temperature at 900°C. The
trend of the HMOR from room temperature up to 900°C may not
be a linear change, as indicated by the lines used to connect the
room temperature and 900°C data points. This trend differed
from the previous studies which showed a sharp increase in MOR
values near the 1100°C temperature.

Figure 10.11 [14] describes the expected influence of
temperature on the MOR stress for a variety of monolithic
refractory materials. These monolithic materials exhibit expected
similar trends in temperature-dependent MOR strengths, as seen in
the previous investigations. The classification of castable types is

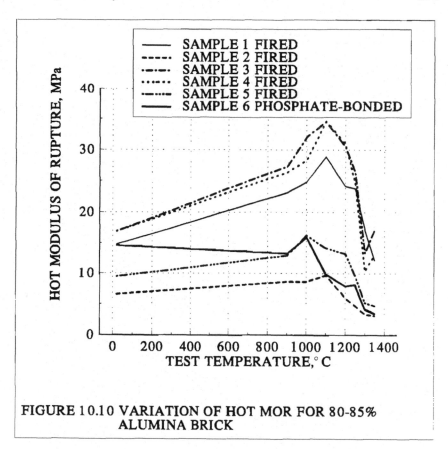

FIGURE 10.10 VARIATION OF HOT MOR FOR 80-85%
ALUMINA BRICK

part of an effort in revising British standards relating to the classification of refractories.

Other investigations [15-20] have also found similar trends in temperature-dependent MOR of castable refractories.

IV. COMPRESSIVE STRENGTH

Just as with the evaluation of the tensile strength, the compressive strength is dependent on several factors. These factors include the dimensionality of the stress state, temperatures and the loading rate.

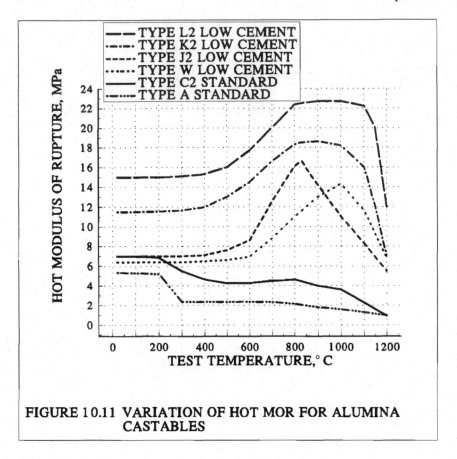

**FIGURE 10.11 VARIATION OF HOT MOR FOR ALUMINA
CASTABLES**

A. The Effect of Multi-Dimensional Stress State
on Ultimate Crushing Stress

Unlike the two-dimensional tensile-tensile stress state, the two-dimensional compressive-compressive stress state increases the magnitude of the compressive stress as compared to the one-dimensional crushing stress f'_c. As shown in Figure 10.1, based on the ratio of the two principal compressive stresses S_1 and S_2, the crushing stress can be 25 to 30% greater than f'_c.

The effect of a three-dimensional compressive stress state has a greater influence on the crushing stress. As described in Figure 10.2, if all three principal compressive stress components S_1, S_2 and

S_3 are kept equal, they can be increased to a level of more than six times f'_c.

It can be concluded that the compressive crushing stress is greatly influenced by the dimensionality of the compressive stress state. The presence of a compressive stress can reduce the tensile failure stress. For a total tensile stress state, the tensile failure stress, as determined from a uniaxial test specimen, is not altered.

B. Effects of Temperature on Hot Crushing Strength

The available literature on the hot crushing strength of refractories is very limited. A considerable amount of work has been conducted on cold crushing strength in which the samples are first heated to a prescribed temperature and then the crushing strength is evaluated at room temperature. By testing heated samples over a range of temperatures, the thermal effects on the cold crushing strength can be determined. However, our interest here is with regard to the crushing strength at higher temperatures. This property is called the hot crushing strength (HCR).

Figures 10.12, 10.13 and 10.14 [2] describe the hot crushing strength behavior for alumina, basic brick and silica brick, respectively. The hot crushing strength of the alumina brick, described in Figure 10.12, shows a decrease in strength with increasing temperature. The 70-72% alumina brick shows a lower hot crushing strength than the 85-86% alumina brick. At about 1700°C, the 70-72% alumina brick reaches a near zero crushing strength, while the 85-86% alumina brick reaches a temporary plateau of strength. Most likely, at higher temperatures, the strength of the 85-86% alumina brick would decrease.

The hot crushing strengths of two different magnesia bricks with different CaO/SiO_2 ratios are shown in Figure 10.13 [21]. Both brick continue to decrease in hot crushing strength as the sample temperature increases. The two basic bricks reverse in maximum strength at about 1250°C.

Figure 10.14 describes the hot crushing strength of two types of silica brick. As shown, both silica bricks maintain a fairly con-

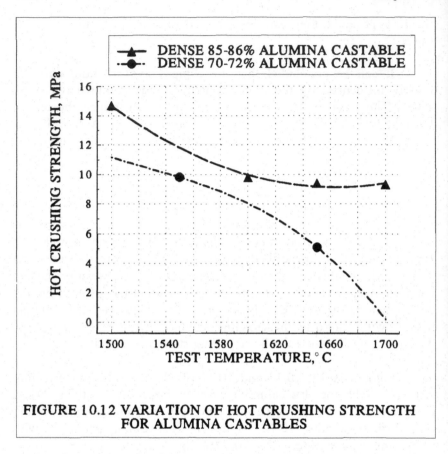

**FIGURE 10.12 VARIATION OF HOT CRUSHING STRENGTH
FOR ALUMINA CASTABLES**

stant strength with increasing temperature. The silica brick with the
1% titano-clay slag addition has a spike in hot strength at 1300°C.

C. Effect of Loading Rate on Hot Crushing Strength

There is very little data in the literature on the influence of
loading rate on hot crushing strength. Figure 10.15 describes the
influence of loading rate on the hot crushing strength of a 27-30%
alumina insulating refractory brick [21]. The faster loading rate is
about 1 MPa per second while the slower loading rate is 1 MPa per
minute. Thus, there is a factor of 60 between the two loading rates.
The reason for the difference in strength is attributed to the
effects of creep. The test specimens subjected to the slower loading

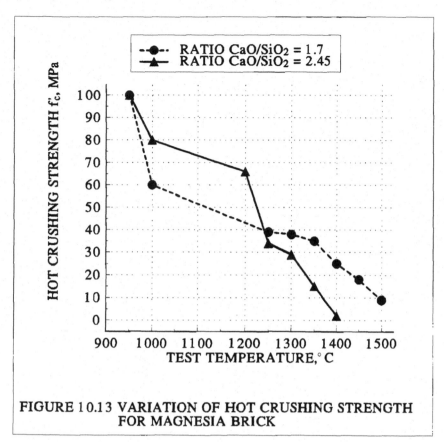

FIGURE 10.13 VARIATION OF HOT CRUSHING STRENGTH FOR MAGNESIA BRICK

rate also accumulate much more creep strain during the test than do the samples with the faster loading rate. Both specimens should fail at the same amount of strain for any specified temperature. Fundamentally, the failure strain is:

$$\epsilon'_f = f'_c / E \qquad (10.1)$$

Equation 10.1 assumes no creep strain has occurred while the sample is loaded. If we include the creep strain (ϵ_c), the equation is revised to:

$$\epsilon'_f = f_c / E + \epsilon_c \qquad (10.2)$$

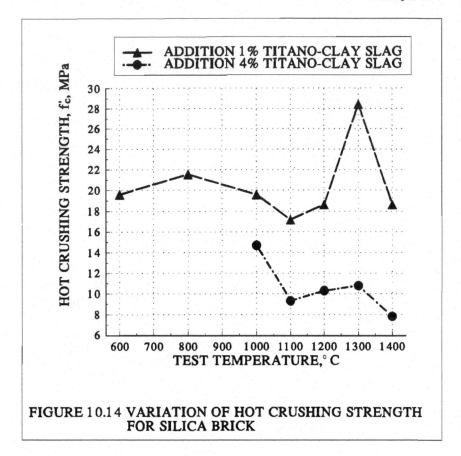

FIGURE 10.14 VARIATION OF HOT CRUSHING STRENGTH FOR SILICA BRICK

If we assume the failure strain is a set quantity, the creep strain will have used a portion of the total available strain. The remaining strain is the elastic-plastic strains developed by the test loading f_c.

Therefore, the slow test ultimate load will be less than the fast test ultimate load. That is:

$$f_c/E < f'_c/E \tag{10.3}$$

or:

$$f_c < f'_c \tag{10.4}$$

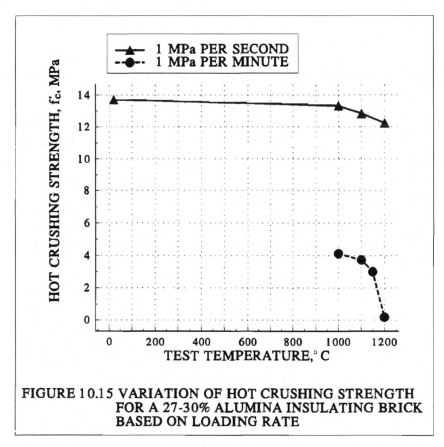

FIGURE 10.15 VARIATION OF HOT CRUSHING STRENGTH
FOR A 27-30% ALUMINA INSULATING BRICK
BASED ON LOADING RATE

As the loading rate is made slower, more creep strain accumulates in the specimen, requiring the loading f_c to be at a lesser value when the specimen fails in compression.

V. SUMMARY

The test data on refractory materials and other similar brittle type materials indicate that the ultimate tensile strength and ultimate crushing strength will vary with regard to the operating environment imposed on the refractory material.

The dimensionality of the tensile stress state (biaxial or triaxial tensile stress states) does not effect the ultimate tensile stress.

A mix of tension and compression stress components tends to reduce the ultimate tensile stress.

The dimensionality of the compressive stress state (biaxial or triaxial compressive stress state) effectively strengthens the refractory in compression.

Increasing temperature tends to cause most fired basic and alumina refractories to reach a peak HMOR at high temperatures. Beyond this peak, the HMOR tends to fall quite rapidly. Silica brick tends to remain stationary with respect to HMOR at increasing temperatures. The firing temperature tends to have a significant influence on the temperature dependent HMOR. As the firing temperature is increased, the peak HMOR, at the higher temperature range, also increases.

The loading rate also has a significant influence on the hot compressive strength (and ultimate tensile strength). For a slow loading rate, a portion of the working strain range is used by the effects of creep leaving less available strain for the elastic and plastic straining. The effects of the loading rate are only applicable to refractories that have significant creep rates. At lower temperatures, the influence of loading should be significantly less. Likewise creep should be significantly less at lower temperatures.

The effects of thermal shock are not included in the previous discussions of this chapter. The above loading rates considered are quite low relative to the rate of loading when discussing thermal shock.

The chemistry and other manufacturing parameters such as firing temperature all greatly influence the strength of refractories. Also, the operating conditions imposed on the refractories which include the loading rates and the dimensionality of the state have a significant influence on the refractory strength. Therefore, both the manufacturer and the user will influence the strength of refractories.

According to more recent work by Darroudi and Lundy [22] on loading rates applied to high alumina refractories, they concluded:

1.   The tensile or fracture strength, at constant loading rates, showed similar trends as those described in Figures 10.7 and 10.8.

2.   The behavior of tensile or fracture strength at different temperatures and varying loading rates was not straightforward. Three trends were observed:

   a.   Strength increased as loading rate increased at low and high temperatures.

   b.   Strength had an inverse relationship with loading rate at intermediate temperature.

   c.   Loading rate had no effect on strength in the regions between intermediate and high temperatures.

These observations appear to differ somewhat from previous investigations.

REFERENCES

1.   Griffith, A. A., The Phenomena of Rupture and Flow in Solids, Phil. Trans. Roy. Soc. (London) A221, 163-198, 1920.

2.   Griffith, A. A., The Theory of Rupture, Proc. 1st Int. Con. Applied Mechanics, Delft, The Netherlands, 1924, pp. 55-63.

3.   Bridgman, P. W., The Effect of Hydrostatic Pressure on the Fracture of Brittle Substances, J. Appl. Phys., Vol. 18, pp. 246-258 (1947).

4.   McClintock, F. A., Friction on Griffith Cracks in Rocks Under Pressure, Proceedings 4th U.S. National Congress on Applied Mechanics, Vol. 2,1962, pp. 1015-1022.

5.   Erdögan, F. and Sih, G. C., On the Crack Extension in Plates Under Plane Loading and Transverse Shear, Trans., ASME, J. Basic Eng. D85, pp. 519-527 (1963).

6.   McClintock, F. A. and Argon, A. S., <u>Mechanical Behavior of Materials,</u> Addison-Wesley Publishing Co., Inc., (Canada),

1966, pp 448-497.

7.   Jaeger, J. C. and Cook, N. G. W., <u>Fundamentals of Rock Mechanics</u>, Second Edition, John Wiley & Sons, Inc., NY, 1976.

8.   Romstad, K. M., Taylor, M. A. and Herrmann, L. R., Numerical Biaxial Characterization for Concrete, ASCE, J. Eng. Mech. Div., Vol. 100, EM5, pp. 935-948 (1974).

9.   Darwin, D. and Pecknold, P. A., Non-Linear Biaxial Stress-Strain Law for Concrete, ASE, J. Eng. Mech. Div., Vol. 103, EM2, pp. 229-241, (1977).

10. Cedolin, L., Crutzen, Y. R. J. and Sandro, D. P., Triaxial Stress-Strain Relationships for Concrete, ASCE, J. Engr. Mech. Div., Vol. 103, EM3, pp. 423-439 (1977).

11. Jackson, B. and Laming, J., The Significance of Mechanical Properties of Basic Refractories at Evaluated Temperatures, Trans. Brit. Cer. Soc., Vol. 68, pp. 21-28 (1969).

12. Miller, E. D. and Davies, B., Modulus of Rupture of Alumina-Silica Refractories at Elevated Temperatures, Cer. Bull. Amer. Cer Soc., Vol. 45, No. 8, pp. 710-713 (1966).

13. Greaves, E. I., The Modulus of Rupture of High-Alumina Refractories at Elevated Temperatures and Their Performance in Arc-Furnace Roofs, Trans. Brit. Cer. Soc., Vol. 68, pp. 15-19 (1969).

14. Padgett, G. C. and Palin, F. T., Test Methods for Monolithic Materials, Advances in Ceramics, Amer. Cer. Soc.,Vol. 13, pp. 81-96 (1985).

15. Clavaud, B., Kiehl, J. P. and Radal, J. P., A New Generation of Low-Cement Castable, Advances in Ceramics, Amer. Cer. Soc., Vol. 13, pp. 274-84 (1985).

16. Banerjee, S., Kilgore, R. V. and Knowlton, D. A., Low-Moisture Castables: Properties and Applications, Advances in Ceramics, Amer. Cer. Soc., Vol. 13, pp. 257-273 (1985).

17. Richmond, C. and Chaille, High Performance Castable for Severe Applications, Advances in Ceramics, Amer. Cer. Soc., Vol. 13, pp 230-256 (1985).

18. Weaver, E. P., Talley, R. W., and Enger, A. J., High Technology Castables, Advances in Ceramics, Amer. Cer. Soc., Vol. 13, pp. 219-244 (1985).

19. Volk, R. J., Engineering Properties of Ultra-Low Cement Castables for Monolithic Ladle Linings, AISE, Iron and Steel Engr. pp. 15-18, Dec., 1987.

20. Kappmeyer, K. K. and Manning, R. H., Evaluating High-Alumina Brick, Amer. Cer. Soc., Ceramic Bulletin, Vol. 42, No. 7, pp. 398-403 (1963).

21. Clavaud, B., Hot Mechanical Properties of Refractories–Compressive Strength, Soc. of European Refractory Prod., Trans., pp. 113-119, Circa 1975.

22. Darroudi, T. and Lundy, R. A., Effects of Temperature and Stressing Rate on Fracture Strength of a Series of High Al_2O_3 Refractories, Ceramic Bulletin, American Ceramic Society, Vol. 66, No. 7, pp. 1139-1143 (1987).

11
Thermal Expansion Data

I. INTRODUCTION

Because many refractory lining systems are exposed to high temperature environments, the need for the thermal expansion data on refractory materials is obvious. As previously discussed, refractory lining systems are typically contained within steel support structures or vessels. In order to evaluate the thermal expansion stresses, one must first conduct a thermal analysis to evaluate the lining and support structure temperatures. The second part of the lining investigation is the actual evaluation of the thermal stresses using the stress-strain data, the creep data and the thermal expansion data. These three types of material properties, along with other assumptions regarding the brick joint behavior and the support structure restraint, are then used to arrive at the refractory lining thermal stresses. The following thermal expansion data are provided based on various investigative results.

II. THERMAL EXPANSION DATA ON
FIRED REFRACTORIES

Figures 11.1 and 11.2 provide a summary of thermal expansion data for a variety of fired refractories [1]. Additional background on the development of these data is found in Reference 2. Investigations continue to evaluate the thermal expansion properties of new refractory chemistries [3,4].

The thermal expansion data are called *reversible,* meaning that the refractory will exhibit this expansion behavior on heatup as well as cooldown. Some of the refractory expansion curves are not straight lines, indicating that the expansion is non-linear and reversible. The silica brick exhibit the greatest non-linear reversible expansion behavior. For refractory materials that exhibit significant non-linear expansion behavior, calculating the coefficient of thermal expansion (α) may require several values over the temperature range of interest. Typically, α is calculated as:

$$\alpha_i = LE_i\%/(T_i - T_R)100 \qquad\qquad 11.1$$

where LE_i is the linear expansion, expressed as a percent for the maximum temperature T_i of interest and T_R is room temperature. α_i is then applicable for the temperature range from T_R to T_i. The α value may change significantly as the temperature of interest increases. Therefore, α becomes temperature-dependent. This α calculating procedure is most applicable for the silica type refractories.

It should be noted that refractories are manufactured using a mix of various materials. It is the chemical mix of the refractories that causes the resulting thermal expansion of the refractory brick. Since the chemistry may vary by a few percent, it is expected that the thermal expansion may vary from lot to lot of a specified type of brick.

Table I describes the calculated coefficients of thermal expansion for a variety of refractory brick. With the exception of the silica brick, α is calculated using the largest temperatures in Figures 11.1 and 11.2. The silica brick is calculated for two

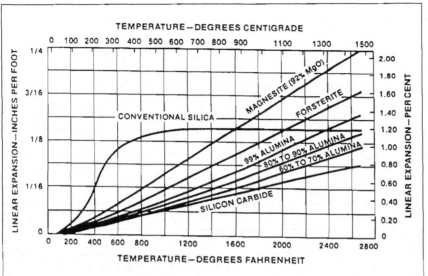

FIGURE 11.1 REVERSIBLE THERMAL EXPANSION DATA FOR FIRED REFRACTORY BRICK [1]

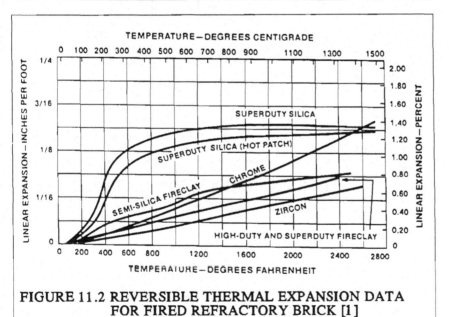

FIGURE 11.2 REVERSIBLE THERMAL EXPANSION DATA FOR FIRED REFRACTORY BRICK [1]

temperatures (300 and 1300°C) to demonstrate the temperature dependency of α for silica brick.

Table I

Estimated Coefficient of Thermal
Expansion for Various Types of
Refractories

Type of Refractory Brick	α mm/mm°C x 10^{-5}	α in./in.°F x 10^{-6}
Fireclay Brick	5.4 to 6.12	3 to 3.4
60 to 70% Alumina	6.3 to 6.84	3.5 to 3.8
80 to 90% Alumina	7.2 to 7.92	4.0 to 4.4
99% Alumina	9.0 to 9.36	5.0 to 5.2
Zircon	4.5 to 4.86	2.5 to 2.7
Chrome	8.82 to 9.18	4.9 to 5.1
Magnesite 92%	14.0 to 14.2	7.8 to 7.9
Conv. Silica at 300°C	30.6 to 31.3	17 to 17.4
at 1300°C	10.1 to 10.6	5.6 to 5.9
Super Silica at 300°C	35.0 to 35.5	19.4 to 19.7
at 1300°C	10.1 to 10.6	5.6 to 5.9
Fired Dolomite	13.3	7.4
Resin Bond Dolomite	11.5	6.4

III. THERMAL EXPANSION DATA ON UNFIRED REFRACTORIES

The thermal expansion coefficients of unfired refractories will vary greatly from those of the previously discussed fired refractory brick. Quite often prior to heatup or curing temperature, dryout occurs causing, in some cases, significant shrinkage. Additional shrinkage occurs at temperatures above 980 to 1100°C due to sintering. Therefore, the user may want to obtain specific thermal expansion data from the manufacturer that includes the shrinkage effects.

REFERENCES

1. Harbison-Walker Handbook of Refractory Practice, Second Edition, Harbison-Walker Refractories Co., Pittsburgh, PA, 1980.

2. Ruh, E. and Wallace, R. W., Thermal Expansion of Refractory Brick, Bulletin, Am. Ceramic Soc., p. 52 (1963).

3. Chen, K.-J., Lee, T.-F., Chang, H.-Y. and Ko, Y.-C., Thermal Expansion of Alumino Silicate Refractory Brick, Ceramics Bulletin, Amer. Cer. Soc., Vol. 61, No. 8, pp. 866-871 (1982).

4. Huang, B. Y. and McGee, T. D., Secondary Expansion of Mullite Refractories Containing Calcined Bauxite and Calcined Clay, Ceramics Bulletin, Amer. Cer. Soc., Vol. 67, No. 7, pp. 1235-1238 (1988).

12
Fundamentals of Refractory Brick Arch Behavior

I. INTRODUCTION

The structural behavior of the stone, or masonry, arch has been a subject of interest for many decades and continues to be of interest to modern engineers. For those who wish to review some of the historical background on masonry arches, see Reference 1. Modern theoretical aspects of arch structural behavior can be found in References 2 and 4. Handbooks can be used to evaluate the forces and displacements of arches in which the various arch equations are condensed into fairly simplified parameters, (see, for example, Ref. 5). However, all modern arch theory assumes the arch is constructed of material which can accept equally both compressive and tensile stresses.

The following discussion attempts to provide the user with information on the basic behavior of circular refractory brick arches. The intention is to equip the user with information necessary to make better decisions with regard to improved refractory arch design.

II. BASICS OF ARCH BEHAVIOR

The primary loading imposed on the refractory brick arch is the thermal expansion loading. One can approximate the arch gravity load stresses from the arch weight using classical circular arch equations [5]. Typically, these stresses are quite small compared to the thermal expansion stresses in the constrained arch. Most refractory brick arch designs, however, accommodate the thermal expansion through various mechanisms. An arch design that has worked successfully in industry is the *sprung arch* design. Figure 12.1 describes the basic parts of the circular sprung arch [6]. The arch horizontal thrust is resisted by the support steel located at each skew.

Figure 12.2 provides additional details of the arch support steel. As shown, each end of a tie bar is connected to two vertical members, called buckstays, which are positioned at each side of the arch and equally spaced along the length of the arch. These buckstays are pinned at floor level and extend upward to the level of the tie bar. A spring is used to connect one end of the tie bar to the adjacent buckstay. The spring stiffness is selected such that the horizontal arch expansion will not significantly increase the horizontal thrust beyond a reasonable level above the arch gravity weight thrust. Figure 12.3 is used here to define the various geometric parts of the arch. Not all of these parameters will be used is discussing arch expansion forces.

Basically, when the arch expands thermally, it transforms from an arch in which all the joints are in full contact or full bearing to a three-hinged arch, as illustrated in Figure 12.4. The arch horizontal thermal growth (dS) is often evaluated using Equation 12.1:

$$dS = \alpha\Delta(S + H) \qquad (12.1)$$

where $S + H$ is the arch span measured to the midthickness locations at each skew, α is the coefficient of thermal expansion of

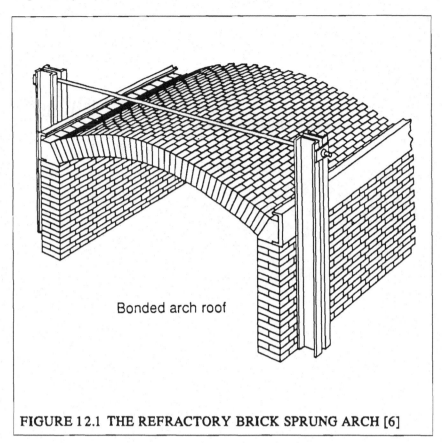

Bonded arch roof

FIGURE 12.1 THE REFRACTORY BRICK SPRUNG ARCH [6]

the arch refractory brick and ΔT is an average increase in temperature of the arch from the installed temperature.

The thermal expansion allowance, for the case of the sprung arch, is accommodated by the adjustment of the nut supporting the spring on the tie bar. Backing off the nut a distance that will accommodate the arch expansion dS results in the heated arch developing expansion forces of similar magnitude to the arch gravity loading. If for some reason the expansion allowance is insufficient (the arch heats to a higher temperature), then the arch hinge condition of Figure 12.4b would develop. The arch would rise upward causing hinges to occur at the bottom side at the arch center (crown) and at the top sides of the two skews. If the arch

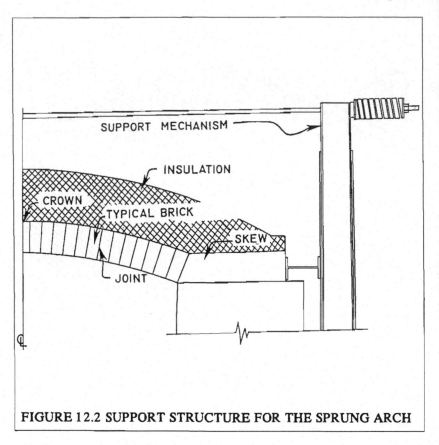

FIGURE 12.2 SUPPORT STRUCTURE FOR THE SPRUNG ARCH

is not heated to the expected temperature, then the arch hinge condition of Figure 12.4c would result. That is, the arch would not rise above the initial installed position. In this case, the hinge at the crown would be at the top side and the two hinges at the skews would be at the bottom side of the two skew joints.

The above description of the arch behavior with respect to arch heatup is sufficient to provide a description of the arch horizontal expansion forces (F_H) developed at each arch skew (see Figure 12.4). Since a three-hinged arch is formed at heatup, the resulting arch forces are statistically determinant. The horizontal thrust, F_H, per foot of arch length described in Figure 12.5 is estimated by [7]:

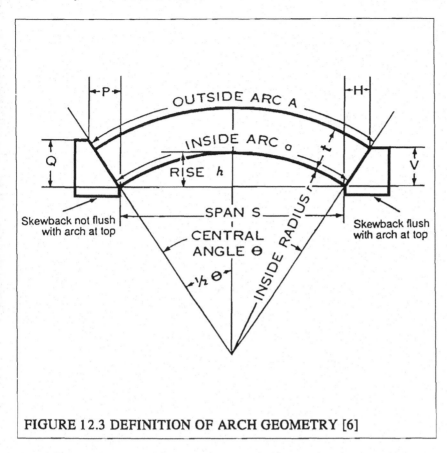

FIGURE 12.3 DEFINITION OF ARCH GEOMETRY [6]

$$F_H \cong WS'/8h' \qquad (12.2)$$

where W is the arch weight per foot of arch length, S' is the span of the two skew hinge points and h' is the vertical distance between the crown and skew hinge points. As illustrated in Figures 12.4b and 12.4c, during the heated condition, the h' value can range from h_{NH} (lower value) to h_{TH} (upper value). As the S' value would be a maximum with h_{NH} and a minimum with h_{TH}. If the arch has a small rise, which most refractory brick arches have, then Equation 12.2 is fairly accurate. For the case in which insufficient or no expansion allowance is used (see Figure 12.4b), the thrust is:

$$F_{TH} = W(S + H)/8\,h_{TH} \qquad (12.3)$$

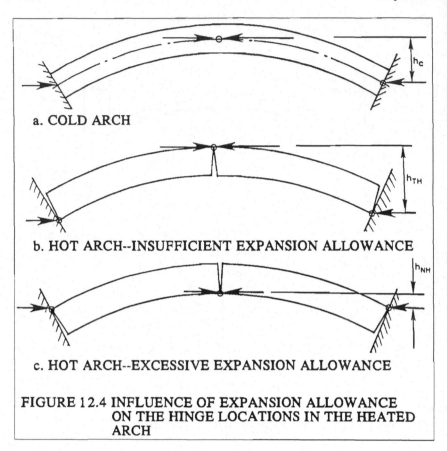

a. COLD ARCH

b. HOT ARCH--INSUFFICIENT EXPANSION ALLOWANCE

c. HOT ARCH--EXCESSIVE EXPANSION ALLOWANCE

FIGURE 12.4 INFLUENCE OF EXPANSION ALLOWANCE
ON THE HINGE LOCATIONS IN THE HEATED
ARCH

For too much expansion allowance (see Figure 12.4c), the thrust is:

$$F_{NH} = WS/8 \, h_{NH} \qquad (12.4)$$

Here, $F_{NH} > F_{TH}$ because $S + H$ is greater than S in the numerator and h_{NH} is smaller than h_{TH} in the denominator. However, the thrust is not an expansion force, but rather a modification of the arch gravity weight. As shown in Equations 12.2 through 12.4, the arch temperature is not used regardless of expansion allowance use. If expansion allowance is not used, the arch will form hinges as shown in Figure 12.4b. The result is a higher gravity load thrust.

McDowell [8] arrives at another form of the equation for the

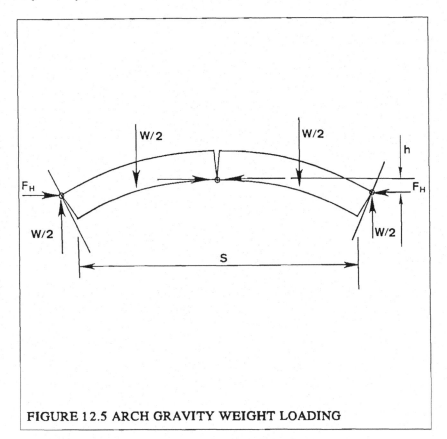

FIGURE 12.5 ARCH GRAVITY WEIGHT LOADING

thrust F_H:

$$F_H = [S/4h - h/3S] \, W/2 \qquad (12.5)$$

Equation 12.2 is a simplified version of Equation 12.5. Establishing a common denominator with the brackets, Equation 12.5 can be rearranged as follows:

$$F_H = [(3S^2 - 4h^2)/12hS] \, W/2 \qquad (12.6)$$

Typically, the arch span (S) values are considerably greater than the arch rise (h) values. In the equation, the numerator $3S^2$ is considerably larger than $4h^2$. Therefore, the absence of $4h^2$ does not

significantly affect the resulting value of F_H. For example, assume
an arch with the following properties:

$$W \;=\; 500 \text{ kg}$$
$$S \;=\; 2235 \text{ mm}$$
$$h \;=\; 240 \text{ mm}$$

Substituting into Equations 12.2 and 12.5:

$$F_H = (500 \times 2235)/(8 \times 240) = 582 \text{ kg}$$

$$F_H = [2235/(4 \times 240) - 240/(3 \times 2235)] \, 250 = 574 \text{ kg}$$

Equation 12.2 predicts an arch thrust about 1.4% higher than
Equation 12.5.

McDowell [7] also provides a limit to the arch thickness (t)
as a function of the rise h, span S and central angle θ of the arch,
expressed as:

$$t \;<\; \{h - [1/(48S)]\}/\cos(\theta/2) \qquad\qquad (12.7)$$

or in a simplified form:

$$t \;<\; [h - (0.02808/S)]/\cos(\theta/2)$$

The constant, 48, is based on a lower limit of the ratio of rise h to
span S, in which a lower limit of 1/4 inch per foot of span, (a ratio
of 1/48) is established. Therefore, this ratio is dimensionless. The
reason for the limitation is to minimize the adverse condition which
arises when in-sufficient or no expansion allowance is used, as
illustrated in Figure 12.4b. As the thickness increases, the vertical
distance h_{NH} between hinges becomes quite small. When this
condition occurs, the arch becomes unstable. That is, the thrust
becomes   very   large (see   Equation   12.4), with   a
decreasing h_{NH}. As the thrust increases, the refractory would
experience either excessive plastic flow or brittle fracture at the
hinges. By keeping the arch thickness at a value less than that of
Equation 12.7, the lower value of h_{NH} is not reached. That is, the
top side of the arch at the skew is lower than a hinge which would

occur at the inside of the arch crown.

Basically, it can be concluded that arches of reasonable rise will form the three hinges upon heatup. If insufficient expansion allowance is used and if the arch is too thick, then the arch becomes unstable due to the position of the exterior two hinges at the skews relative to the interior hinge at the crown. If a slight excess of expansion allowance is used, then the arch is quite stable during heatup due to the large h_{TH}. It should also be mentioned that extreme expansion allowance could be undesirable. In this case, the arch could also collapse inward when the unheated arch hinges (exterior at crown and interior at skews) are at a similar level and h_{TH} approaches zero.

III. ARCH THERMAL DISPLACEMENTS

In the previous section, the arch thrust load and the expected range of this thrust load was defined. The basis of the definition of the arch thrust was that the arch formed hinges during heatup. The locations of the hinges (either inside or outside of the arch) are based on the amount of expansion allowance relative to the arch expansion. The following discussion addresses the manner in which the arch expands when exposed to operating temperatures. The purpose here is to better define the arch thermal expansion such that an imposed expansion allowance can be used.

The refractory brick arch is typically exposed to a temperature environment in which the inside of the arch is at a higher temperature than the outside of the arch. Referring to Figure 12.6a, the cold arch is exposed to a linear through-thickness temperature distribution in which the inside temperature (hotface), T_h, is greater than the outside (coldface) temperature, T_c. The average temperature (T_{AVE}) of the arch is:

$$T_{AVE} = (T_h + T_c)/2 \qquad (12.8)$$

Therefore, the thermal growth of the arch span at the inside of the arch is illustrated in Figure 12.6b and defined as:

$$dS' = \alpha(T_{AVE} - T_R)S \qquad (12.9)$$

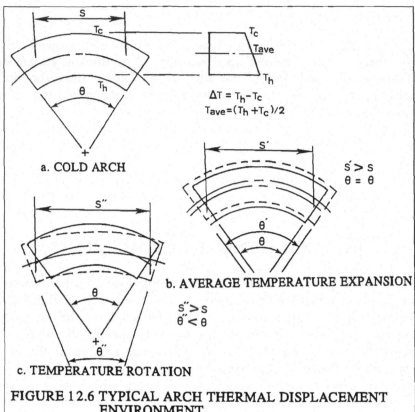

a. COLD ARCH

$\Delta T = T_h - T_c$

$T_{ave} = (T_h + T_c)/2$

$s' > s$
$\theta = \theta$

b. AVERAGE TEMPERATURE EXPANSION

$s'' > s$
$\theta'' < \theta$

c. TEMPERATURE ROTATION

FIGURE 12.6 TYPICAL ARCH THERMAL DISPLACEMENT ENVIRONMENT

or:

$$dS' = \alpha \Delta TS$$

where T_R is the installed or room temperature. dS' is defined as thermal expansion that occurs due to the average increase in arch temperature.

The second portion of the arch expansion is that part which occurs due to the through-thickness temperature gradient. As illustrated in Figure 12.6c, the arch rotates causing the central angle to increase from the original installed central angle of θ to a lesser central angle of θ'':

$$\theta'' < \theta \tag{12.10}$$

Note that the central angle from the average increase in temperature remains equal to the original central angle:

$$\theta'' = \theta \tag{12.11}$$

The change in the central angle is defined as:

$$\phi = \theta'' - \theta \tag{12.12}$$

The angle ϕ can also be defined in terms of the temperature difference, dT, through the arch:

$$dT = T_H - T_c \tag{12.13}$$

The unit thermal growth at either the inside or outside of the arch is:

$$dL = dT\alpha$$

The unit angle of rotation is:

$$d\phi = \tan^{-1}(2dT\alpha/t) \tag{12.14}$$

where $d\phi$ is the angular change for a unit length of arch, as measured along the centerline of the arch. The total angular change is ϕ, as defined in Equation 12.11. ϕ is calculated as:

$$\phi = Ld\phi \tag{12.15}$$

where L is the arch length along the arch centerline, as defined by (see Figure 12.3 for arch geometry):

$$L = [2\pi(r + t/2)]\theta/360 \tag{12.16}$$

Therefore:

$$\phi = \tan^{-1}(2dT\alpha/t)[2\pi(r + t/2)]\theta/360 \tag{12.17}$$

Simplifying:

$$\phi = \tan^{-1}(2dT\alpha/t)[0.0175(r + t/2)\theta] \qquad (12.18)$$

The revised arch angle θ'' is:

$$\theta'' = \theta - \phi \qquad (12.19)$$

The revised arch radius r'' is taken as the inverse of the central angle change:

$$r'' = (\theta/\theta'')r \qquad (12.20)$$

The revised span S'' is:

$$S'' = 2r'' \sin \theta'' \qquad (12.21)$$

Based on the through thickness temperature difference, the additional span length dS'' is:

$$dS'' = S'' - S \qquad (12.22)$$

Therefore, the total increase in the arch span dS_{TOT} is:

$$dS_{TOT} = dS' + dS'' \qquad (12.23)$$

If the through thickness temperature difference is small, then the dS'' will be small. However, the effects of temperature difference can add a significant amount of arch span expansion.

   In summary, the evaluation of the arch span expansion should include both the average temperature effects and the through-thickness temperature difference effects, resulting in an improved estimate of the required expansion allowance for the refractory brick arch.

IV. SUMMARY

   The refractory arch is typically exposed to a temperature environment that is characterized as a higher hotface temperature than coldface temperature. Because of the arch thermal expansion, a geometry change takes place. This new expanded geometry does not fit within the installed skew position because of two primary parts of the expansion. The arch increases in size due to the average increase in temperature and rotates due to the through thickness temperature difference. Even if the skews are displaced horizontally to match the increase in span, the slope of the skew face does not match the new central angle of the arch. As a result, the arch will form three hinges, one at the center crown location and one at each skew face. The three hinges formed within the arch are due to mismatching between the expanded arch and the position of the installed skews and the slope of the skew face.

Expansion allowance is used to accommodate the expanded arch geometry. The skew can be displaced horizontally to allow for the increase in the arch span. However, the skew is not usually rotated to accommodate the change in the central angle of the arch. A slight excess in expansion allowance usually creates a more stable heated arch than when insufficient expansion allowance is used.

Insufficient expansion allowance coupled with too thick of an arch, as defined by McDowell, will result in an unstable arch. Too much expansion allowance will also create an unstable arch and arch collapse.

In the design of an arch, the following items should be addressed:

1. The operating temperature distribution of the arch should be fully understood in order to calculate expansion allowance.

2. The appropriate arch thickness should be used.

3. The appropriate expansion allowance should be incorporated into the arch design.

4. The highest stresses occur at the three hinges, and the three hinges are at the crown and skew faces. The refractory brick with the highest structural integrity should be chosen for these locations.

5.   Since the arch thermal loading reverts to an arch gravity weight loading, the arch loading is a stress controlled loading. The ultimate crushing strength is a part of the strength criteria used to select the best refractory brick for the arch.

The flat arch has been gaining considerable popularity in the industrial community. No extensive investigative work has been found in the literature on the flat arch. As a result, no discussion has been included here. It is expected that the success of this arch is based on the support system design surrounding it. It is expected that the flat arch and its support structure are designed to accept a thermal expansion load and that these loadings may be somewhat greater than the previously discussed circular arch.

The catenary arch is another form of refractory brick arch used in industry. The disadvantage of the catenary arch is the vertical rise of this arch. Typically, the rise of the catenary arch is much greater than the circular arch. The result is that a greater space is required for this arch. Most likely, the catenary arch will form hinges similar to that of the circular arch. However, the catenary arch should be much more stable than the circular arch. Also, expansion allowance may not be required for the catenary arch meaning that the skews can remain fixed at the installed position.

REFERENCES

1.   Heyman, J., The Masonry Arch, Ellis Horwood Ltd., Halsted Press, John Wiley & Sons, First Edition, 1982.

2.   Flügge, W., Stresses in Shells, Springer-Verlag, New York, 1966.

3.   Timoshenko, S. and Woinowsky-Krieger, S., Theory of Plates and Shells, Second Edition, McGraw Hill Book Company, 1959.

4.   Wang, C.-K., Statistically Indeterminate Structures, McGraw-Hill Book Company, New York, 1953.

5. Roark, R. J. and Young, W. C., <u>Formulas for Stress and Strain</u>, Fifth Edition, McGraw Hill Book Company, 1975.

6. Harbison-Walker Handbook of Refractory Practices, First Edition, Harbison-Walker Refractories, One Gateway Center, Pittsburgh, PA 15222, 1992.

7. Schacht, C. A., Stress-Strain and Other Thermomechanical Environments Imposed on Refractory Systems in Steelmaking Operations, Amer. Cer. Soc., Ceramic Trans., Advances in Refractory Technology, Vol. 4, (1990).

8. McDowell, J. S., Sprung-Arch Roofs for High Temperature Furnaces, Blast Furnace and Steel Plant, Part III, Blast Furnace and Steel Plant, February, 1940.

13
Fundamentals of Brick-Lined Cylindrical Shells

I. INTRODUCTION

Of all the various shapes of industrial process vessels, the cylindrical vessel geometry appears to be the most predominant. The objective of this chapter is to describe the thermomechanical behavior of the cylindrical refractory-lined vessel. Typically, the vessel shell is a structural or pressure-vessel grade steel. In the following discussion, it is also assumed that the refractory lining is made up of refractory brick.

The purpose of the refractory-lined process vessel is usually to contain high-temperature process materials. These materials may be granular, liquid or gas.

The second function of the refractory lining is to isolate the support structure (usually made of steel) from the process temperature. In most cases, structural and pressure-vessel grade steel cannot be exposed to temperatures much above 350 to 450°C.

The purpose of the steel support structure is to provide the necessary tensile strength to constrain the lining in the heated condition. To accomplish this, the lining must insulate the steel structure from excessive temperatures.

The third function of the refractory lining is to control heat loss from the process. Heat loss control may be economic (due to high energy costs) or because the process requires the maintenance of a stable temperature environment for product quality control.

II. BACKGROUND

The cylindrical refractory-lined process vessel is fundamentally exposed to a combination of operating pressures and temperatures. The pressure and temperature process environment is applied to the inside (hotface side) of the lining, as illustrated in Figure 13.1. In most cases the lining consists of several layers of varying quality refractories. The interior layer of refractory lining that is exposed to the process is often called the *working lining*. In some vessels that contain molten liquids, such as molten steel, the linings between the shell and working lining are often referred to as the *safety* or *insulating linings*. Sometimes the lining behind the working lining is called the *back-up lining*. Other terms given to lining systems include *wear skin* for the working lining and *safety skin* for the safety lining. Regardless of the terms used, the lining system serves the same purpose: to contain the process environment.

As previously indicated, the objective of this chapter is to provide insight into the thermomechanical behavior of the refractory lining system. To accomplish this task a simplified numerical example is used to portray the fundamental aspects of lining mechanical behavior and to show how the lining interacts with the vessel shell. It should also be mentioned that the hand calculation procedure was chosen since this procedure allows separation of all the interacting parts of the lining behavior. This would not be possible in a finite element analysis procedure. Several assumptions are made to reduce the number of considerations to only those necessary to understand the lining behavior and to make the hand calculations of this numerical example more tractable.

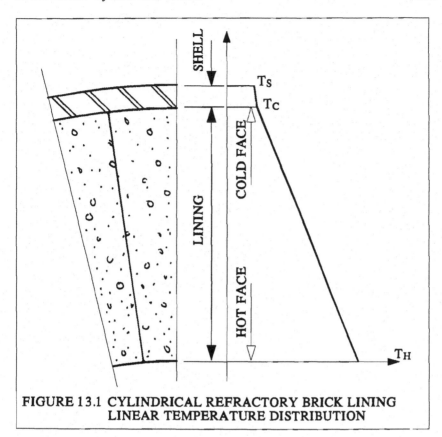

FIGURE 13.1 CYLINDRICAL REFRACTORY BRICK LINING LINEAR TEMPERATURE DISTRIBUTION

The first assumption used in the example is that the refractory material is linearly elastic. That is, no inelastic flow (plastic or creep) occurs. Simplifying the refractory material to elastic behavior reduces the calculations to reasonable limits.

The second assumption is that the temperature gradient will be linear through the thickness. Therefore, we are evaluating what could be termed the steady-state operating temperature. During transient heatup of a lining system, the through-thickness gradient becomes non-linear. Attempting to use hand calculations with transient heatup temperature would greatly complicate this effort.

The third assumption is that the brick joints can not support any tensile loading. This is realistic for a dry joint. For a mortar joint, a small magnitude of tensile loading can be supported. However, when comparing the low tensile strength of most mortar joints to the much greater compressive strength of the brick lining, the assumption of zero mortar joint strength does not have a significant influence on the accuracy of the solution. It will be shown that the assumption of non-tension in the mortar joint does not severely jeopardize the predicting of lining behavior.

The fourth assumption is perhaps, the most important. A lining thickness was chosen such that the vessel shell will be at a low temperature, causing the shell to expand less in the radial direction than the refractory lining. Therefore, the shell will constrain the lining and induce compressive stresses in the lining. The vessel shell will, in turn, develop a tensile stress to equilibrate the lining compressive stress. The lining compressive stress serves an important role in that a mechanical seal is formed in the brick joints thereby preventing penetration of process materials. If the lining is made too thin, the higher shell temperature will cause the shell to expand radially more than the lining. As a result, the lining will not be constrained by the vessel shell, resulting in a loose lining.

The fifth assumption is that no expansion allowance is taken into consideration. The joints are assumed to be ideal rigid interfaces.

III. DEFINING EQUATIONS FOR THE NUMERICAL EXAMPLE

The basic behavior of a cylindrical lining, exposed to a through-thickness linear temperature gradient and restrained by the shell, is described in Figure 13.1. Typically, as illustrated in Figure 13.2, the inner portion (or hotface side) of the refractory lining develops a compressive thermal stress and the shell develops an equilibrating tensile stress. Radial cracks develop on the cold side of the lining joints as the refractory has little or no tensile strength. A radial pressure (P) develops between the refractory lining and the vessel shell due to the thermal restraint imposed on the lining by the shell (see Figure 13.3).

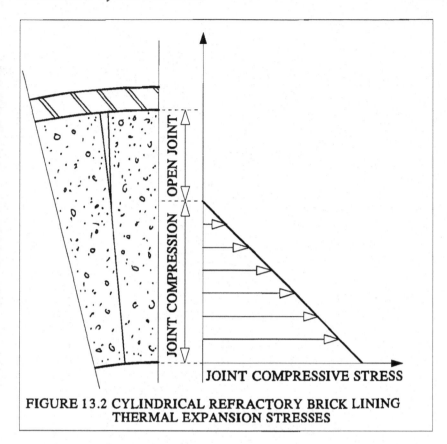

JOINT COMPRESSIVE STRESS

FIGURE 13.2 CYLINDRICAL REFRACTORY BRICK LINING
THERMAL EXPANSION STRESSES

Figure 13.3 describes the various parameters of the cylindrical refractory lining geometry that are used in the following equations. As noted, the lining joint is not in contact on the cold side of the lining. The following example will demonstrate how the joint condition is developed. There are two components of the lining linear temperature that will be dealt with in the example problem. They are:

1. The average temperature of the lining thickness under consideration. This is the portion of the temperature distribution that creates the expansion forces. The lining expansion force (F_L) will be equilibrated by the shell expansion

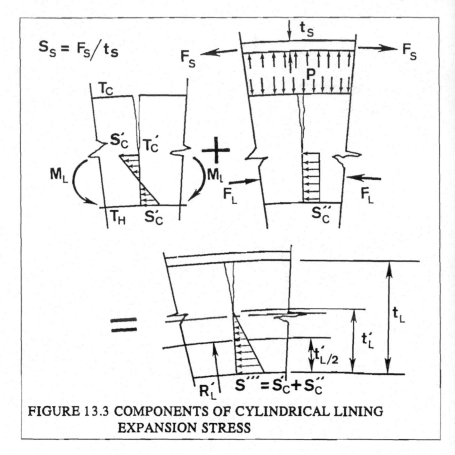

FIGURE 13.3 COMPONENTS OF CYLINDRICAL LINING
EXPANSION STRESS

force F_S.   The lining expansion creates a
uniform stress S_C''.

2.   The second component is the gradient portion of
the temperature distribution in the lining. This
portion of the temperature distribution creates a
thermal moment (M_L) for the portion of the
lining under consideration. This portion of the
temperature distribution does not have any
influence on the expansion forces F_L or F_S.
However, the thermal moment does create
thermal stresses in the lining. As shown in
Figure 13.3, the moment stresses are defined as

S_c'. S_c' is the compressive stress on the hotface side of the lining and adds to the compressive stress S_c''. On the coldface side, for the lining thickness under consideration, S'_c is the tensile stress and subtracts from the compressive stress S_c''.

As shown in Figure 13.3, the addition of the two components results in the final compressive stress state S_c'''.

A. Defining the Equations

The basic behavior of the cylindrical vessel lining system can be demonstrated by using the following vessel equations. The hoop (circumferential) stress equation for a pressurized cylindrical vessel shell is defined as:

$$S_C = PR/T \tag{13.1}$$

where:

S_c	=	Hoop (circumferential) stress
R	=	Vessel radius
t	=	Vessel shell thickness
P	=	Radial pressure

Equation 13.1 assumes the ratio of the vessel shell radius to the vessel shell greater than about 10. That is, thin shell vessel behavior is assumed.

Dividing both sides of the equation by the modulus of elasticity (E), the left side of the equation becomes the hoop strain ϵ_c:

$$\epsilon_C = PR/tE \tag{13.2}$$

The incremental change in the circumferential length (ΔC) of the vessel wall is obtained by multiplying the circumferential strain by the circumferential length ($C = 2\pi R$):

$$2\pi R\epsilon_c = \Delta C = 2\pi PR^2/tE \tag{13.3}$$

The radial displacement (ΔR) can be determined knowing $\Delta C = 2\pi \Delta R$. Substituting:

$$\Delta R = PR^2/tE \qquad (13.4)$$

Equation 13.4 forms the basis for our evaluation of the vessel lining system. The vessel lining is defined to consist of two parts, the vessel lining and the vessel shell. As previously discussed, it will be assumed that the lining was selected of sufficient thickness such that a low shell temperature is achieved. This also means that the shell expands radially less than the lining. As a result, a radial restraint pressure (P) will exist between the inside face of the vessel wall and the coldface side of the lining (see Figure 13.3). The previous equations (Equations 13.1-13.4) are now redefined for each of the two components (the vessel lining and vessel shell).

The radial interference pressure (P) is of equal magnitude on the shell and lining. However, this pressure causes inward radial displacement of the lining defined as:

$$\Delta R_L = PR'^2_L/t'_L E_L \qquad (13.5)$$

where:

R'_L = Effective midthickness radius of lining
t'_L = Effective working thickness of lining
E_L = MOE of lining

Referring to Figure 13.3, the full lining thickness is t_L. However, as we will demonstrate in the following example, the full lining thickness will not be active in thermal expansion interaction between the lining and shell.

Because of the restraint loading on the vessel shell, the vessel shell undergoes an outward radial displacement:

$$\Delta R_s = PR^2_s/t_s E_s \qquad (13.6)$$

where:

R_s = Radius of shell
t_s = Thickness of shell
E_s = Elastic modulus of shell

The sum of the inward lining displacement and the outward shell radial displacement ($\Delta\delta_P$) is defined as:

$$\Delta\delta_P = P(R'^2_L/t'_L E_L + R_s^2/t_s E_s) \qquad (13.7)$$

Turning now to the thermal expansion of the lining and the shell, the radial expansion of the vessel lining (ΔR_{TL}) is:

$$\Delta R_{TL} = \alpha_L R'_L \Delta T_L \qquad (13.8)$$

where α_L is the coefficient of thermal expansion of the lining and ΔT_L is the midthickness temperature increase (at R'_L) as measured from room temperature.

The vessel shell radial expansion (ΔR_{TS}) is:

$$\Delta R_{TS} = \alpha_S R_S \Delta T_S \qquad (13.9)$$

where α_S is the coefficient of thermal expansion of the vessel shell steel and ΔT_S is the increase in temperature of the vessel shell as measured from room temperature. Since the lining will expand radially more than the shell, the resulting difference in the radial thermal growth is the thermal interference ($\Delta\delta_T$), defined as:

$$\Delta\delta_T = \alpha_L R'_L \Delta T_L - \alpha_S R_S \Delta T_S \qquad (13.10)$$

To satisfy both the force and displacement conditions for the thermal interference between the lining and shell:

$$\Delta\delta_T = \Delta\delta_P \qquad (13.11)$$

Substituting Equation 13.7 into Equation 13.11:

$$\Delta\delta_T = P[R'^2_L/t'_L E_L + R^2_s/t_s E_s] \qquad (13.12)$$

An average elastic modulus is used for the refractory working lining. Although the approach is approximate, the fundamental thermomechanical behavior of the lining and shell can be identified. In actual linings, the elastic modulus varies as a function of temperature, and on the hotface side of the lining significant plastic flow occurs. The *non-tension* joint loading criterion is employed using trial and error solutions of the preceding equations.

To add more accuracy to this approximate approach, the lining can be divided into several thin cylindrical parts. Equation 13.12 would then be redefined as:

$$\Delta\delta_T = P[\, \Sigma^n_{i=1}\, (R'^2_L/t'_L E_L)_i + R^2_s/t_s E_s\,] \qquad (13.13)$$

where the n represents the number of thinner cylindrical lining sections. The lining parameters within the parentheses are for each of the thinner sections. Equation 13.13 may be more applicable if the MOE varies significantly due to the effects of temperature. It should be re-emphasized that these equations are still very approximate compared to more sophisticated finite element methods.

IV. EXAMPLE OF CYLINDRICAL LINING BEHAVIOR

The cylindrical vessel is assumed to be lined with refractory that has the following properties:

α_L = 6.3 x 10^{-6} mm/mm°C
t_L = 229 mm (full lining thickness, $t'_L = t_L$)
R'_L = 1638 mm (R_{inside} = 1524 mm, $R_{outside}$ = 1753 mm)
E_L = 17.4 GPa

Let us assume a steady-state heat transfer analysis was conducted and the following temperatures were obtained:

T_H = 1093°C
T_C = 200°C

For the vessel shell:

$$\alpha_s = 11.7 \times 10^{-6} \, \text{mm/mm}^\circ\text{C}$$
$$t_s = 32 \, \text{mm}$$
$$R_s = 1768 \, \text{mm}$$
$$E_s = 207 \, \text{GPa}$$
$$\Delta T_s = 204 - 21 = 183^\circ\text{C}$$

Trial Solution No. 1:

Assuming the full lining thickness will develop the lining expansion stresses, the average increase in the lining temperature from ambient is:

$$\Delta T_L = \{[(1093 + 200)/2] - 21\} = 626^\circ\text{C}$$

Substituting the parameters into Equation 13.10, the thermal interference is:

$$\Delta\delta_T = (6.3 \times 10^{-6})(1753)(626) - (11.7 \times 10^{-6})(1768)(183)$$

$$= 6.91 - 3.79 = 3.12 \, \text{mm}$$

Therefore, the lining will grow radially more than the shell. This satisfies our condition that an expansion restraint must be imposed by the shell to provide a stable lining.

Substituting the various parameters into Equation 13.12:

$$\Delta\delta_T = P\{[1638^2/(229)(17.4)] + [1768^2/(32)(207)]\}$$

$$= P(673 + 472) = P(1145)$$

Solving for the radial thermal interference pressure (compressive pressure, P):

$$P = \Delta\delta_T/1145 = 3.12/1145 = 0.0027 \, \text{GPa} = 2.7 \, \text{MPa}$$

Substituting into Equation 13.1, the vessel shell circumferential tensile expansion stress is:

$$S_S = F_S/t_S = PR_S/t_S = (2.7)(1728)/32 = 146 \text{ MPa}$$

The circumferential compressive stress S''_C (see Figure 13.3) in the refractory lining is:

$$S''_C = -(2.7)(1638)/229 = -19.3 \text{ MPa}$$

Note that S''_C represents the average compressive stress in the lining and not the influence of the temperature gradient.

Since the refractory has a through-thickness temperature gradient, these gradient stresses (S'_C; see Figure 13.3) must be added to the average stress, S''_C. Therefore, at the outside surface (against the shell) of the lining:

$$S'_C = +[E_L \alpha_L (T_H - T_C)]/2$$

$$= +(17.4)(6.3 \times 10^{-6})(1093 - 200)/2$$

$$S'_C = +0.0489 \text{ GPa} = 48.9 \text{ MPa}$$

The total stress (S''') on the cold side of the working lining is:

$$S''' = S'_C + S''_C = +48.9 - 19.3$$

$$= +29.6 \text{ MPa}$$

Since this tensile stress is clearly in excess of the tensile strength of mortared brick joints, a second trial is required using only a portion of the lining thickness on the hotface side of the vessel lining, illustrated in Figure 13.3.

Trial Solution No. 2:

Assume the inside two-thirds of the lining will develop the full expansion force, therefore:

$$t'_L = (2/3)229 = 153 \text{ mm}$$

T_C = 500°C (at 2/3 location)

ΔT_L = [(1093 + 500)/2 - 21] = 756

ΔT_s = 183°C, identical to trial 1

Substituting into Equation 13.10:

$$\Delta\delta_T = (6.3 \times 10^{-6})(1600)(756) - 3.79$$

$$= 3.83 \text{ mm}$$

Substituting into Equation 13.12:

$$\Delta\delta_T = P[1600^2/(153)(17.4) + 472] = P[1434]$$

Solving for the radial thermal interference pressure (compressive pressure P):

$$P = \Delta\delta_T/1434 = 3.83/1434 = 2.67 \text{ MPa}$$

The vessel shell circumferential tensile stress (S_C; see Figure 13.3) is:

$$S_s = + (2.67)(1768)/32 = +148 \text{ MPa}$$

The average circumferential compressive stress (S''_C; see Figure 13.3) in the refractory lining is:

$$S''_C = - (2.67)(1600)/153 = - 27.92 \text{ MPa}$$

The circumferential lining stress at the interior cold face (at 153 mm from lining hotface) due to the temperature gradient is:

$$S'_C = + (17.4)(6.3 \times 10^{-6})(1093 - 500)/2$$

$$= + 0.0325 \text{ GPa} = 32.50 \text{ MPa}$$

The lining stress S''' at the location 153 mm from the hotface is:

$$S''' = + 32.50 - 27.92 = 4.58 \text{ MPa}$$

$$S''' = + 32.50 - 27.92 = 4.58 \text{ MPa}$$

Likewise, the circumferential lining stress at the hotface is:

$$S''' = -32.50 - 27.92 = -60.42 \text{ MPa}$$

Trial 2 is an improved solution over Trial 1. The interior tensile stress in the joint is still significant. The tensile stress of 4.58 MPa at the interior location is still a bit too high. For a dry joint, the stress should be zero. Perhaps a third trial is required using a t' of about 130 mm.

The example problem demonstrates:

1. The coldface side of the lining can develop radial tension cracks (open radial joints).

2. The lack of circumferential tension stress in the lining results in the vessel shell equilibrating all of the lining compressive expansion stress.

3. Only the average temperature of the lining causes the expansion restraint with the vessel shell. The gradient portion of the through-thickness temperature distribution only adds to the lining thermal stress.

4. The condition of no tension stress in the lining joints results in the center of gravity of the lining compressive expansion stress shifting toward the hotface of the lining.

5. The lining should be installed with thin mortar joints and the linings must be installed tight against the vessel shell. In some instances the loss of the expansion interference can be attributed to (1) a lack of tightness in lining installation; (2) the use of crushable material behind the working lining; or (3) the use of too thick a compressible blanket or board material, which can result in a

loose working lining and subsequent penetration of process materials such as molten liquids.

6.   The insulating lining must have sufficient crushing strength. The example shows a compressive stress (P) of about -2.67 MPa applied to insulating brick used between the working lining and the shell. If rapid heatup is used, then higher expansion forces will be imposed on the insulating brick, as described in Figure 13.4.

It should be noted that the analytical method used in the example is approximate and is limited to elastic/non-tension joints. A more sophisticated analytical method such as a finite element

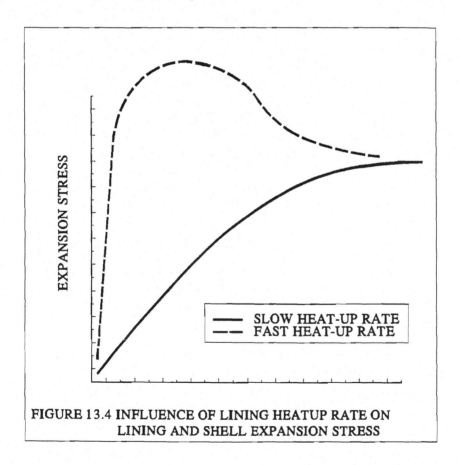

FIGURE 13.4 INFLUENCE OF LINING HEATUP RATE ON
LINING AND SHELL EXPANSION STRESS

analysis is required if the lining's elastic modulus is temperature dependent, if the temperature distribution is transient (non-linear) or if non-linear plastic flow is to be included in the hotface region of the lining.

V. EXPANSION ALLOWANCE CONSIDERATIONS

Unlike the circular refractory arch, the cylindrical refractory lined vessel does not form hinges when subjected to thermal expansion. The expansion stress-strain values continue to increase with increasing temperature. Therefore, in some instances where the refractory lining material has a combination of high stiffness, high coefficient of thermal expansion, a highly restraining support structure and high operating temperatures, or is exposed to high heatup rates, expansion allowance may be a necessary part of the lining design.

The mechanism used for expansion allowance comes in several different forms. These mechanisms include the common mortar joint, mortar joint sheeting material (for example, cardboard) and the use of compressive blanket material. The present state of expansion allowance technology is very approximate. The following information hopefully provides a more technical basis for incorporating expansion allowance into the lining system that requires some relief from thermal expansion.

A. Mortar Joints

The use of mortar joints, or at least the taking into account the effects of the mortar joints [1], for expansion allowance can greatly assist in determining if other expansion allowance mechanisms are necessary.

Perhaps the first initial insight into the appreciation of the influence of the potential expansion forces can be determined from the following equation that roughly estimates the stiffness of the lining relative to the stiffness of the vessel shell, defined as:

$$ST_{L/S} = E_L t_L / E_s t_s \qquad (13.14)$$

This equation is evaluated at various operating temperatures to estimate roughly the influence of the lining in creating significant expansion stresses in the vessel shell. If the ratio approaches 1, then the stiffness of the lining is significant and expansion stresses should be evaluated. As shown in Section III, the estimate of the lining thickness, t'_L, is a critical part of this evaluation. The finite element model of linings has the capability to evaluate the effective thickness of the lining automatically as part of the model predictions on lining behavior.

The mortar joint is a critical part of the evaluation of the expansion allowance evaluation. We have previously demonstrated the magnitude of the circumferential and radial stresses which exist in the lining (see Section III). The ratio between the hoop stress (S''_c) and the radial stress (P) is a function of the lining radius and lining thickness. That is, from Equation 13.1:

$$S''_c = (R'_L/t'_L)P$$

(13.15)

The hoop stress is greater than the radial stress by the ratio of the lining radius to the lining thickness. It is possible, due to lining geometry, that the hoop stress can be ten times greater than the radial stress. Therefore, the radial joints are exposed to a much higher stress than the circumferential joints. The result is that radial joints are compressed and will undergo a greater loss of thickness than the circumferential joint.

The second aspect of the lining joints is the number of joints in the two directions. Typically, the number of circumferential joints is limited to two or three, based on the number of lining components. The number of radial joints is considerably greater. In a considerable number of cylindrical refractory-lined vessels, the ratio ($J_{R/C}$) of the number of radial joints (N_R) to the number of circumferential joints (N_C) is:

$$J_{R/C} = N_R / N_C$$

(13.16)

Since the radial joints are exposed to the higher hoop stress, the ratio ($S_{R/C}$) of hoop to radial stress is:

$$S_{R/C} = R'_L/t'_L \qquad (13.17)$$

Chapter 8, on mortar joint behavior, demonstrated that the greater the stress imposed on mortar joints the, greater the joint collapse. Combining the effects on the number of joints with the magnitude of compressive loading on the respective joints, the influence of the radial joints relative to the circumferential joints in developing expansion allowance is estimated as:

$$R_{EXPALL} = J_{R/C} \, S_{R/C} \qquad (13.18)$$

or, redefined, as:

$$R_{EXPALL} = N_R \, R'_L / N_C \, t'_L \qquad (13.19)$$

This is an extremely rough estimate on the influence of radial joints, but does indicate that radial joints have a more significant role in creating expansion allowance than circumferential joints.

B. Use of Compressible Insulating Blankets

In some instances, the mortar joints do not provide sufficient expansion allowance. As a result, other expansion allowance mechanisms must be incorporated into the lining design. Addressed here will be the use of compressible insulating blankets. The key to the successful use of insulating blankets is understanding the compressive deformation behavior of these types of materials.

Figure 13.5 describes the compressive stress-displacement relationship for an insulating blanket that has an initial (unloaded) density of 160 to 192 kg/m^3 [2]. The lighter weight materials exhibit considerable displacement at very low compressive loadings. At about 3 to 4 MPa, the blanket displaces about 85% of its original thickness. That is, the thickness of the blanket at 3 to 4 MPa is about 15% of its original unloaded thickness. At higher compressive loadings, above 4 MPa, the thickness is reduced by only about another 6 to 8%. A compressive load of about 65 MPa

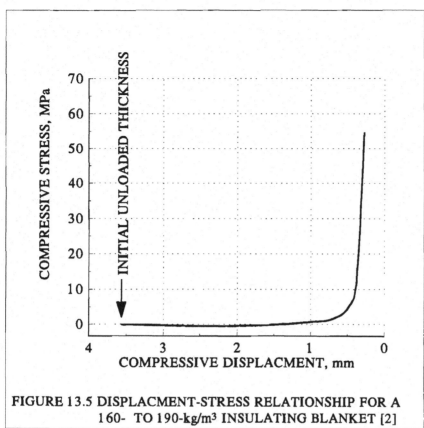

FIGURE 13.5 DISPLACMENT-STRESS RELATIONSHIP FOR A
160- TO 190-kg/m³ INSULATING BLANKET [2]

causes the blanket to displace about 93% of its original thickness.
The previous example indicated a radial compressive loading (P) of
about 3 MPa. Therefore, if an insulating blanket has a density of
160 to 192 kg/m³, then the blanket will compress about 85%.

For example, a 160- to 812-kg/m³ density insulating blanket
is chosen to be used as an expansion allowance mechanism in a
cylindrical refractory-lined vessel. The blanket would be placed
circumferentially against the shell. The desired expansion
allowance is 5 mm. The desired thickness of blanket to be used
would be 6 mm (5/0.85 = 5.88).

Figure 13.6 describes the compressive stress-displacement

relationship for an 800-kg/m^3 density insulating blanket [2]. Because of this greater density, considerably greater compressive loading is required to compress this material.   In this case, a compressive stress of 3 to 4 MPa would only displace this blanket by about 35%. For extreme compressive loadings of 65 MPa, this blanket would displace about 60%. Therefore, for radial loadings from refractory, this blanket is expected to displace about 35% of its thickness. That is, this material would compress such that the compressed thickness would be 65% of its original thickness. If a 6-mm thick blanket is used against the shell, the expected expansion allowance from this material is about 2 mm (0.35 x 6 = 2.1).   Looking at this blanket selection from the reverse direction, if a desired expansion allowance of, say, 5 mm is desired, then a thickness of about 14 mm (5/0.35 = 14.3) should be used.

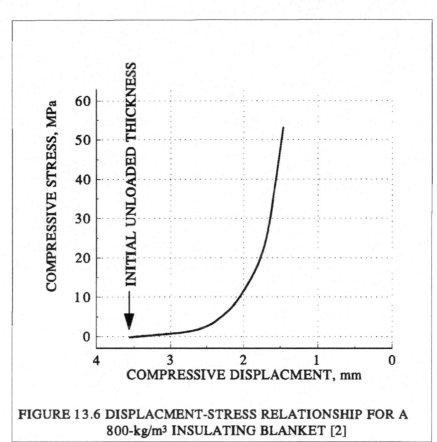

FIGURE 13.6 DISPLACMENT-STRESS RELATIONSHIP FOR A
800-kg/m^3 INSULATING BLANKET [2]

The compressive stress-displacement data [2] indicates a significant relationship to the density of the blanket. Figure 13.7 provides an interpretation of the compressive stress-displacement data for a range of expected insulating blanket densities [3]. As indicated, the expected displacement (or percent of thickness that is compressed, C) of the blanket material as a function of the density is estimated as:

$$C = 100 - 0.1Q_B \qquad (13.20)$$

where C is the percent of displaced thickness, due to a compressive stress in the range of 2 to 5 MPa, and the insulating blanket density is in kg/m³. The constant of 0.1 is an estimated constant for this relationship. Using Equation 13.15, a blanket density of

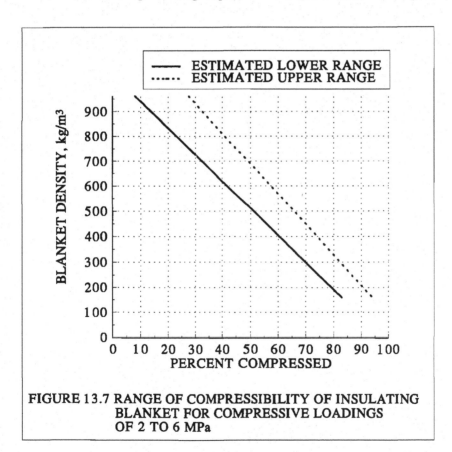

FIGURE 13.7 RANGE OF COMPRESSIBILITY OF INSULATING BLANKET FOR COMPRESSIVE LOADINGS OF 2 TO 6 MPa

20% [100 - 0.1(800) = 20%]. Therefore, if a blanket thickness of 12 mm is chosen, this blanket is expected to displace about 2.5 mm (0.20 x 12 = 2.4) when subjected to typical radial loadings from a cylindrical refractory lining.

It should be emphasized that insulating blanket materials may have a maximum operating temperature suggested or recommended by the manufacturer. Temperatures above the maximum operating temperature may cause blanket shrinkage or other forms of blanket deterioration. Therefore, a greater expansion would occur in excess of the expected expansion allowance. If this occurs, the lining may lose the necessary restraining force provided by the shell. This, in turn, would cause a loss of the mechanical seal in the lining joints and an undesirable penetration of process materials into the lining.

Another problem associated with the use of insulating blanket materials occurs when lighter density insulating blankets are used in process vessels that are exposed to continuous cyclic flexing of the vessel shell. This flexing can cause a working of the rigid lining brick against the softer insulating blanket, resulting in mechanical wear and deterioration of the blanket material. This then causes an unexpected increase in the expansion allowance.

When using an insulating blanket as an expansion allowance material, the heat transfer analysis should include the insulating effects of this blanket. If the blanket is severely compressed, it may lose some of its insulating capacity because of the densification that occurs.

VI. THIN CYLINDRICAL REFRACTORY LININGS

Quite often in certain industries, thin refractory linings are used because of the lesser need for the insulating capacity of the refractory materials. In these instances, the process temperatures are in the intermediate to lower range (less than about 650°C). The supporting vessel shell can be exposed to temperatures up to 400°C without any significant loss in structural integrity. As a result, thin refractory linings are used. In many instances, such as

result, thin refractory linings are used. In many instances, such as in the refinery industry, the refractory linings are monolithic (castable or gunned type) linings.

When the linings become too thin, a highly undesirable condition occurs that can cause lining deterioration. This condition is the radial thermal growth of the vessel shell relative to the thermal growth of the vessel lining. Because of the lesser insulating capacity of the thin lining, the shell temperature is relatively higher than in a thick lining system. As a result, the radial thermal growth of the vessel shell is greater than the radial thermal growth of the lining. When this occurs, the lining is not constrained by the shell.

The condition just described is demonstrated by the following example. Assume a cylindrical vessel lined with a 100-mm thick castable lining is placed in a 1200-mm diameter vessel. The vessel shell is 25 mm thick. The vessel shell is a typical pressure-vessel grade carbon steel. Since the lining system temperatures are a critical part of this investigation, the thermal conductivity (K) is also defined. The material properties of the carbon steel used in this investigation are:

$$\alpha_S = 11.7 \times 10^{-6} \text{ m/m}^\circ\text{C}$$

$$K_S = 35 \text{ W/mK}$$

The castable material has an Al_2O_3 content of about 58% and SiO_2 content of about 32%. The material properties to be used in this investigation are:

$$\alpha_L = 3.8 \times 10^{-6} \text{ m/m}^\circ\text{C}$$

$$K_L = 1.30 \text{ W/mK}$$

The process temperature is 630°C. The shell temperature is calculated using the classical heat transfer equation in most heat transfer text books [see, for example, Ref. 4], and defined as:

$$Q = \frac{T_H - T_R}{\dfrac{1}{h_H} + \displaystyle\sum_{i=1}^{n} \dfrac{t_i}{K_i} + \dfrac{1}{h_C}} \tag{13.21}$$

where T_H is the process temperature, T_R is the exterior ambient temperature, h_H is the hotface heat transfer coefficient, h_C is the vessel shell exterior surface heat transfer coefficient, t_i is the thickness of a lining material component and K_i is the corresponding thermal conductivity of this lining material component. The following values are used for the heat transfer coefficient:

$$h_H = 28.5 \ W/m^2K$$

$$h_C = 11.5 \ W/m^2K$$

The exterior ambient temperature is set at 22°C. The thicknesses are converted to meter units. Substituting the values into Equation 13.16:

$$Q = \frac{630 - 22}{\dfrac{1}{28.5} + \dfrac{0.10}{1.3} + \dfrac{0.025}{35} + \dfrac{1}{11.5}}$$

$$= \frac{608}{0.035 + 0.077 + 0.00071 + 0.087}$$

The first step of the heat flow calculations are provided to show that the steel shell (third value in denominator) has very little significance on the heat flow. The effect of the steel shell could have been neglected without any significant loss in the accuracy of the solution.

The heat flow is:

$$Q = 3034 \ W/m^2$$

The shell temperature is:

$$T_s - 22 = 0.087(3034)$$

$$T_s = 286°C$$

Because of the low thermal resistance of the vessel shell steel, it can be assumed that the inside of the shell will also be at 286°C.

It can be determined if a restraint condition is established by comparing the radial expansion of the steel shell to the radial expansion of the lining hotface. The lining hotface would be the portion of the lining that would develop the greatest radial expansion. The radial expansion of the vessel shell is:

$$\Delta R_s = \alpha_s R_s \Delta T_s$$

$$= (11.7 \times 10^{-6})(600 \text{ mm})(286 - 22)$$

$$= 1.85 \text{ mm}$$

The radial expansion of the lining hotface is:

$$\Delta R_L = (3.8 \times 10^{-6})(600 - 100)(630 - 22)$$

$$= 1.16 \text{ mm}$$

Since:

$$\Delta R_s > \Delta R_L$$

If the radial expansion of lining thickness is also included using the average lining temperature:

$$\Delta t_L = (3.8 \times 10^{-6})(100)[(630 + 286)/2 - 22]$$

$$= 0.166 \text{ mm}$$

The lining radial expansion is revised to:

$$\Delta R_L = 1.16 + 0.166 = 1.33 \text{ mm}$$

However, the shell expansion is still greater.

The gap between the lining and shell at operating temperatures is:

$$GAP = \Delta R_S - \Delta R_L$$

$$GAP = 1.85 - 1.33 = 0.53 \text{ mm}$$

There is no restraint imposed on the lining by the shell. As a result, the lining will remain loose within the vessel shell during normal operating conditions. Since this is a castable lining, most likely some shrinkage has occurred increasing the gap between the lining and shell.

The need for a low shell temperature is an important consideration when designing a lining system for a low-temperature environment. By reducing the shell temperature through the use of either a thicker lining or using an insulating layer against the shell, the shell expansion is reduced. With high shell temperatures, the lining is subjected to high tensile stresses, resulting in tensile cracking of the lining during operating conditions. If lower shell temperatures are designed into the lining/shell system, the lining is subjected to stabilizing compressive stresses during operating conditions. Since the castable materials are typically stronger in compression than tension, the lower shell temperatures will result in fewer tensile cracks in the lining during operating conditions.

VII. SUMMARY

The results of the numerical examples illustrate several interesting behaviors of cylindrical lined vessels. With regard to the thick lining required for high-temperature operating environments:

1.   The cold side of the lining system cannot develop circumferential compressive expansion stresses. Typically, most refractory brick has coefficients of thermal expansion (α_L) that are typically less than

those of the steel shell (α_s):

$$\alpha_L < \alpha_s$$

Assuming the lining adjacent the shell is at the same temperature ($T_L = T_s$) the circumferential thermal growth of the adjacent lining will not be as great as the circumferential thermal growth of the shell. Therefore, the joints adjacent to the shell must open for the lining to remain tight against the shell.

2. One key factor in creating a lining restraint condition is the thermal resistance of the lining. As the lining increases in thickness (t_L) and/or as the thermal conductivity (K) increases, the restraint forces will also increase. That is, the shell will decrease in temperature as K and t_L increase, causing a shell restraint to occur.

3. A second factor which assists in creating a lining restraint is the lining coefficient of thermal expansion (α_L). As α_L increases, the lining restraint also increases.

4. Some refractory materials have a high α, resulting in large restraint forces. In some instances, the restraint becomes too great, causing either fracturing of the lining material or distortions of the steel shell. In this case, use of expansion allowance is required to reduce the adverse effects of excessive thermal expansion forces.

5. There is a critical balance in the use of expansion allowance materials. Excessive expansion allowance can result in an undesirably loose lining, while an insufficient amount can cause lining deterioration.

With regard to the lining required for lower temperature

operating environments:

1. Typically, the lining is made thin based solely on heat transfer results. That is, the shell temperature is not excessive, therefore, the lining can be thin.

2. Based on thermomechanical (thermal expansion behavior) considerations, the thin lining causes excessive tensile stresses in the lining as a result of greater radial thermal displacements in the vessel shell.

3. The lining thickness of a cylindrical refractory-lined vessel subjected to lower operating temperatures should be evaluated based on thermomechanical considerations.

REFERENCES

1. Schacht, C. A., Fundamental Considerations in the Structural Evaluation of Refractory-Lined Cylindrical Shells, Iron and Steel Engineer, AISE, Vol. 59, No. 6, pp. 44-47 (1982).

2. Cortney, R. L., Compressibility of Expansion Paper, Technical Memorandum, Kaiser Aluminum & Chemical Corp. (National Refractories and Minerals Corporation) Proj. No. 49750, March 11, 1981.

3. Schacht, C. A., AISE Ladle Study SC-9-1, AISE Interim Reports No. 1 through No. 4 and Final Report, September 1984.

14
Fundamentals of Brick Dome Behavior

I. INTRODUCTION

Another geometric form of refractory lining structure seen in industry is the spherical lining. The spherical refractory lining dome is used for large-diameter furnaces which require a complete refractory lining without interior supports. Typically, only the cylindrical wall lining supports the dome around the periphery of the dome base.

Figure 14.1 illustrates the basic parts of the spherical refractory lining structure. The purpose of this chapter is to describe the basic forces and moments in this type of lining when subjected to thermal expansion displacements. Typically, this type of refractory lining is supported by a vessel shell as described in Figure 14.1. The vessel shell expands thermally less than the refractory dome. As a result, hinges occur within the dome structure because of the restraint imposed by the vessel shell. Since the shell restraint is at the skew, the hinges occur in the region of

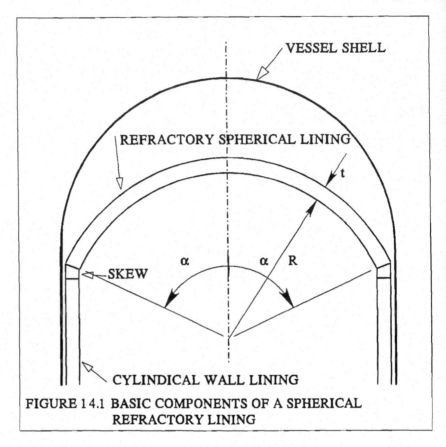

FIGURE 14.1 BASIC COMPONENTS OF A SPHERICAL
             REFRACTORY LINING

the skew. The following discussion will provide insight into the nature of dome behavior in the region of the skew and the locations of the hinges.

II. BACKGROUND

The spherical refractory lining is often designed with a primary working lining, as shown in Figure 14.1. Just as with cylindrical refractory-lined vessels, safety or insulating linings are used to maintain the vessel shell temperatures at reasonable levels. However, when this is done, the shell will thermally expand less than the refractory lining. The result is that the vessel shell restrains the lining from growing to the full potential thermal displacements.

Figure 14.2 describes and illustrates the effect of the vessel restraint at the skew of the spherical lining. Typically, the skew restraint is the only portion of the spherical lining that is restrained by the vessel shell. Sufficient space is provided between the top side of the insulation and the spherical shell such that the upward thermal growth of the dome will not be restrained by the spherical portion of the vessel shell. Figure 14.2 describes the radial thermal growth of the dome and skew as dR_D. The cylindrical portion of the vessel shell restrains the radial growth of the skew to a lesser amount, defined as dR_S. Therefore, the dome is restrained by the amount of:

$$\text{Dome Restraint} = dR_D - dR_S$$

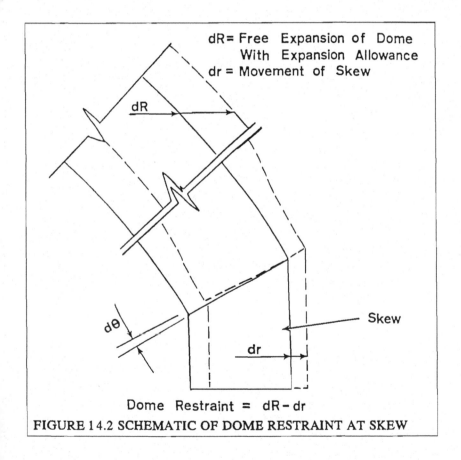

dR = Free Expansion of Dome
      With Expansion Allowance
dr = Movement of Skew

dR

dθ

Skew

dr

Dome Restraint = dR – dr

FIGURE 14.2 SCHEMATIC OF DOME RESTRAINT AT SKEW

The following section discusses the theoretical behavior of spherical shells constructed of materials (such as steel) which resist both tensile and compressive strength. The refractory lining system cannot resist tensile loadings, as previously discussed. However, by reviewing this theoretical behavior, a better understanding on the behavior of spherical refractory linings is provided.

III. THEORETICAL BEHAVIOR OF SPHERICAL SHELLS

The following discussion addresses the theoretical behavior of spherical domes constructed of materials that have strength to resist both tensile and compressive loadings.

The coordinate system used to define the dome is shown in Figure 14.3. Also shown are the shell moments and forces. The N_ϕ force and bending moment M_ϕ are in the meridional direction, while the N_θ force and bending moment M_θ are in the circumferential direction. Spherical shells with a pinned edge (Figure 14.4a) and a fixed edge (Figure 14.4b) are used to illustrate the edge effects on the shell forces and moments. The loading imposed on the shells is an inward pressure loading applied to the exterior side of the shells.

The force and moment behavior is described in Figure 14.5 [1]. As suggested in Figure 14.2, our interest is in the N_ϕ and M_ϕ force and moment. The circumferential hinges in the spherical refractory lining would result from this force and moment. For the hinge-edged dome, the M_ϕ would be zero at the edge. The stresses were calculated assuming [1]:

> t = 143 mm
> R = 60 mm
> v = 0.20
> α = 39°
> Pressure Load, P = 2 MPa

The results show that at the edge (α = 39°), the only stress is a compressive stress (N_ϕ) of about 7 MPa. Of particular interest is the trend of the combined N_ϕ force and M_ϕ moment stresses at the inside and outside surfaces of the dome. The combined maxi-

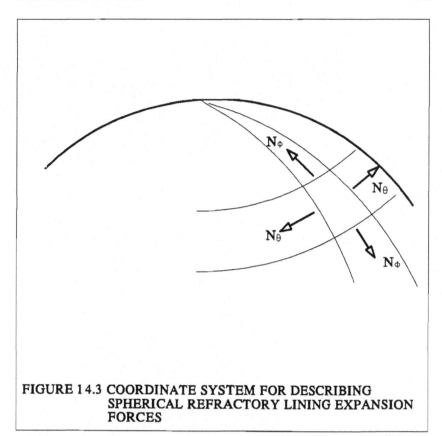

FIGURE 14.3 COORDINATE SYSTEM FOR DESCRIBING
SPHERICAL REFRACTORY LINING EXPANSION
FORCES

mum compressive stress would be at the outside surface (at about α = 30°), and the combined maximum tensile ϕ stress would be at the inside surface (at about α = 32°). The maximum stress is predominantly a result of the bending moment M_ϕ. The M_ϕ maximum, therefore, occurs at a location of α = 30 to 32°. This maximum M_ϕ moment location should be kept in mind when examining the spherical refractory lining behavior later in this chapter.

The second spherical shell has a fixed edge, as illustrated in Figure 14.4b. This dome has the identical geometry as the pinned-edge dome. The stress results of the fixed-edge dome analysis is detailed in Figure 14.6. These results show that the greatest com-

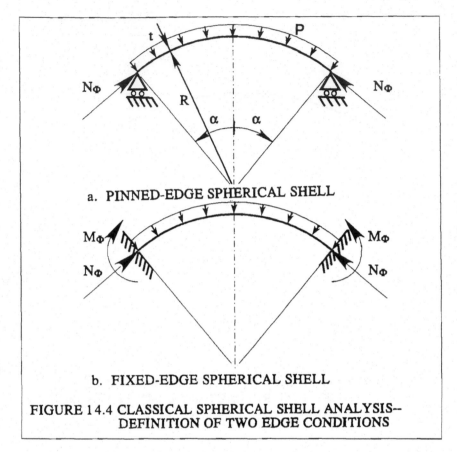

a. PINNED-EDGE SPHERICAL SHELL

b. FIXED-EDGE SPHERICAL SHELL

FIGURE 14.4 CLASSICAL SPHERICAL SHELL ANALYSIS--
DEFINITION OF TWO EDGE CONDITIONS

bined ϕ stresses are at the fixed edge (α = 39°). The second maximum ϕ stress occurs at a location of about α = 24 to 26°. With the fixed condition, the maximum combined ϕ stresses are considerably less than for the pinned-edge condition. These two locations should also be kept in mind when discussing the spherical refractory lining later in this chapter.

The third example is a fixed edge dome [1], but was analyzed by less rigorous and more approximate theoretical equations. This dome had the following geometry:

t = 76 mm
R = 2286 mm

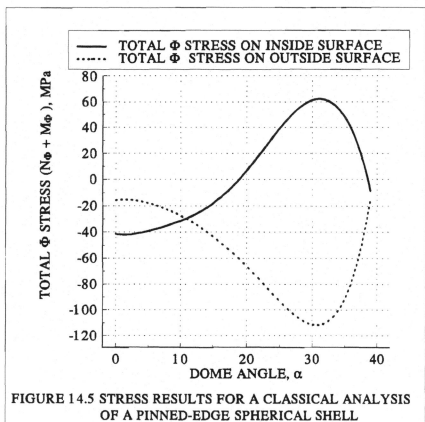

FIGURE 14.5 STRESS RESULTS FOR A CLASSICAL ANALYSIS
OF A PINNED-EDGE SPHERICAL SHELL

$\nu = 0.20$
$\alpha = 35°$
$P = 0.007$ MPa

The stress results are described in Figure 14.7. These results show
the maximum combined ϕ stress is at the fixed edge ($\alpha = 35°$)
and at an interior location of $\alpha = 21$ to $23°$. As for the previous
two examples, the α locations of maximum stress should be kept in
mind.

There are other sources for evaluating the stresses in spherical
shells [2]. Most likely, shell structures with other geometrical and
loading variations would exhibit similar trends in the regions of

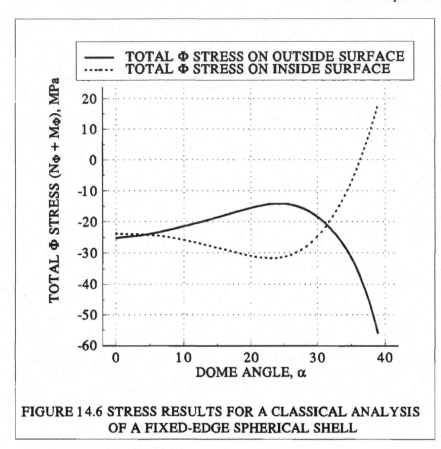

FIGURE 14.6 STRESS RESULTS FOR A CLASSICAL ANALYSIS
OF A FIXED-EDGE SPHERICAL SHELL

maximum φ stress.

The results of the theoretical analyses for spherical shells have shown that the edge condition has a significant influence on the maximum φ stress. Also, the maximum φ stress occurs at the edge for fixed edges. For a fixed-edge dome, a second interior maximum φ stress occurs at an α angle closer to the edge. In all cases, the stresses decay to much lower uniform stresses at the interior part of the dome shell.

The results of the theoretical analyses serve as an important basis for the evaluation of spherical refractory linings. Since refractory lining cannot develop tensile stresses, it will behave

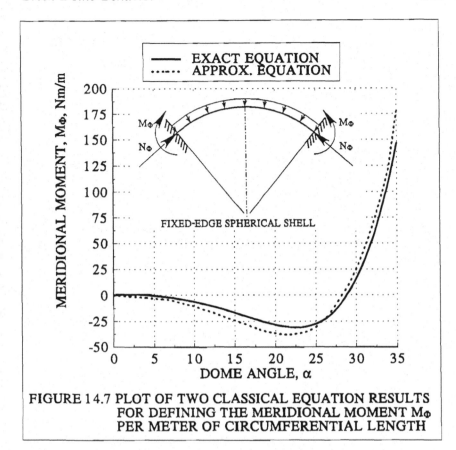

FIGURE 14.7 PLOT OF TWO CLASSICAL EQUATION RESULTS
FOR DEFINING THE MERIDIONAL MOMENT M_Φ
PER METER OF CIRCUMFERENTIAL LENGTH

differently. That is, where maximum moments occur, hinges will most likely develop in the refractory lining. The following section briefly presents the analytical results of a spherical refractory lining and how the resulting behavior is realistic based on the previous theoretical results.

IV. RESULTS OF A FINITE ELEMENT ANALYSIS ON A SPHERICAL REFRACTORY LINING

Several spherical refractory linings have been evaluated, and all have shown a similar behavior in the region of the skew. In all cases, the skew was restrained from complete radial thermal expansion, as previously illustrated in Figure 14.2. The following discussion details the results of finite element analysis results for the refractory spherical lining described in Figure 14.8 [3]. As shown, the lining has the following geometry:

t = 305 mm
R = 4724 mm
v = 0.20
α = 66°

The dome's work lining was constructed of silica brick. Insulating brick were placed on the coldface side of the work lining. Silica brick was used because of the high-temperature working environment imposed on the dome. Since silica brick has a high coefficient of thermal expansion, especially in the lower temperature

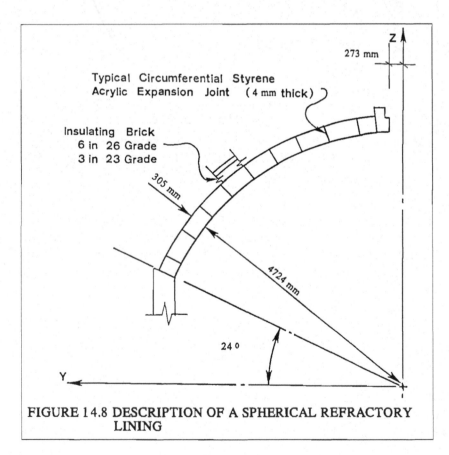

FIGURE 14.8 DESCRIPTION OF A SPHERICAL REFRACTORY
            LINING

range, expansion allowance was used. The expansion allowance was in the form of plastic sheets about 4 mm thick placed at intervals in the circumferential and meridional dome work-lining joints. This plastic would burn out at a few hundred degrees centigrade. Not all of the expansion was accounted for in the expansion allowance. Sufficient expansion allowance was used such that excessive thermal stresses were not developed in the silica work lining. However, the resulting radial expansion of the dome at the skew region was still greater than the skew radial expansion. The skew radial expansion was limited by the cylindrical part of the vessel shell. The vessel cylindrical shell at the coldface side of the insulating brick behind the skew was heated to temperatures that caused a restraint to radial thermal growth of the skew.

The finite element model analysis for the previously described spherical refractory work lining is shown in Figure 14.9. This model was constructed of three-dimensional solid elements, with six elements through the work lining thickness. The insulating brick loading was simulated by applying an inward radial pressure on the top of the dome. The model also included special elements at the joints that would resist compression and separate when subjected to tension. Therefore, hinge locations would be predicted by the model. These special elements were also used to accommodate the expansion allowance at every third circumferential joint. These elements were used to accommodate the meridional joint behavior.

The thermal displacement behavior, as predicted by the model, is shown in Figure 14.10. The displacement results describe a most interesting behavior. Recalling the theoretical results of the previous section, the greatest moments were developed at the edge of the spherical shell. The spherical refractory lining behaves in a very similar manner. The refractory dome experiences two circumferential hinges, the first at the skew face and the second at a short distance up from the skew face.

The hinge at the skew face exhibits a joint opening on the hotface side of the dome, while the second hinge opens the joint on the coldface side of the dome. Both hinges are necessary to accommodate the radial thermal displacement of the dome, which is

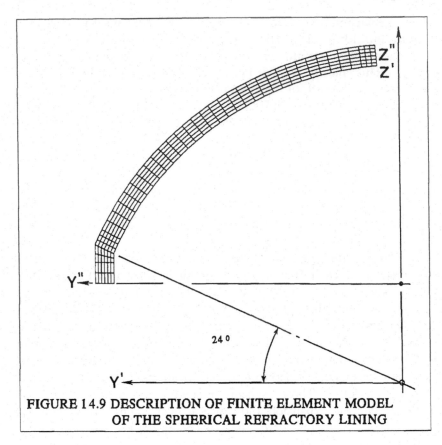

**FIGURE 14.9 DESCRIPTION OF FINITE ELEMENT MODEL
OF THE SPHERICAL REFRACTORY LINING**

greater than the skew. The location of the hinges are also in very favorable agreement with the two maximum moment regions predicted by the classical theoretical spherical shell behavior.

V. SUMMARY

In summary, the refractory spherical lining develops two circumferential hinges, one at the skew face and one a short distance up from the skew. The hinges are formed due to the greater radial displacement of the dome than those of the skew. The hinges are formed to accommodate the differences in the radial

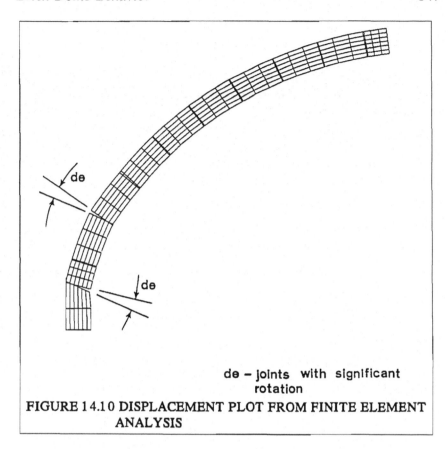

de – joints with significant
rotation

FIGURE 14.10 DISPLACEMENT PLOT FROM FINITE ELEMENT
ANALYSIS

thermal expansion at the skew. These results imply that the stresses in the dome are greatest in the skew region and that better quality refractory brick should be placed in this region of the dome.

The circumferential thermal stresses are greatest at the skew due to the shell restraint at the skew. Expansion allowance can be in the form of that discussed in Chapter 13 on cylindrical lining systems.

The circumferential stresses in the dome are greatest in the region just above the first hinge at the skew. Since a greater radial displacement occurs in the region at the second hinge, the circumferential stresses are greatly reduced.

The circumferential stresses are less in the dome because much less restraint is imposed on the dome.

The use of expansion allowance serves to reduce the radial expansion of the dome, thereby reducing the amount of rotation at each of the two hinges.

In some dome designs, special knuckle joints have been successfully used, as shown in Figure 14.11, to accommodate the rotation at the anticipated hinge locations.

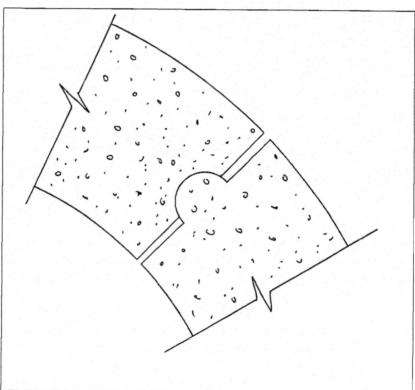

FIGURE 14.11 CIRCUMFERENTIAL HINGE FOR SPHERICAL REFRACTORY LINING

REFERENCES

1. Timoshenko, S. and Woinowsky-Krieger, S., <u>Theory of Plates and Shells</u>, Second Edition, McGraw-Hill Book Company, 1959, pp. 533-558.

2. Roark, R. J. and Young, W. C., <u>Formulas for Stress and Strain</u>, Fifth Edition, McGraw-Hill Book Company, 1975, p. 474.

3. Schacht, C. A., Stress-Strain and Other Thermomechanical Environments Imposed on Refractory Systems In Steelmaking Operations, American Ceramic Society, Cer. Trans., Advances in Refractory Technology, Vol. 4, pp. 205-242 (1990).

15
Fundamentals of Flat Brick Linings

I. INTRODUCTION

Square or rectangular brick-lined structures are used in a wide variety of process vessels. These flat-walled linings can result in highly undesirable distortions based on the features of the design and the design loadings. Flat brick linings are, in some instances, made a part of cylindrical or spherical refractory lining systems. The objective of this chapter is to provide an understanding of the flat lining behavior when subjected to typical temperature and restraint environments.

II. BACKGROUND

This chapter is concerned with the thermomechanical behavior of a flat brick lining during typical operating conditions. It will be assumed that the refractory lining is exposed to a process temperature and a process pressure. The pressure as addressed here comes from the hydrostatic pressure effect of hot liquids within the vessel. Liquids such as molten salt, molten aluminum and molten steel will develop pressure on the hotface of the refractory lining. Based on the magnitude of this pressure, the pressure can have a significant influence on the behavior of the flat brick lining.

The flat brick lining will most likely have a decreasing temperature through the lining thickness. The slope of this decreasing temperature distribution is a function of the magnitude of the process temperature, the thermal material properties of the refractory material, the amount of insulating brick, the magnitude of the outer ambient temperature and other related factors. In the following discussions, only linear through-thickness temperature distributions will be addressed.

With the flat brick linings, two kinds of process loads are addressed: the pressure (stress-controlled) load and the temperature (strain-controlled) load. Both of these loadings have a significant influence on the behavior of the flat brick lining.

III. INFLUENCE OF PROCESS LOADINGS ON FLAT BRICK LINING BEHAVIOR

As previously indicated, both the pressure loading and the thermal loading have a significant influence on the behavior of the flat brick lining. Because the lining is flat, both of these loadings, if applied separately, will cause a distinct lining displacement. In the following discussions, a unit depth of lining is assumed.

A. Pressure Load Behavior

The pressure loading is assumed to be applied to a single span of a flat brick lining. The single span is assumed to be pinned, supported at each end of the span. Figure 15.1a describes the basic single span, flat wall lining and the resulting displacement, D_P, due to the pressure load, P, defined as:

$$D_P = 5PL^4/384 \ EI \qquad (15.1)$$

where L is the span length, E is the modulus of elasticity and I is the moment of inertia (see Figure 15.1a). Since the lining is a brick

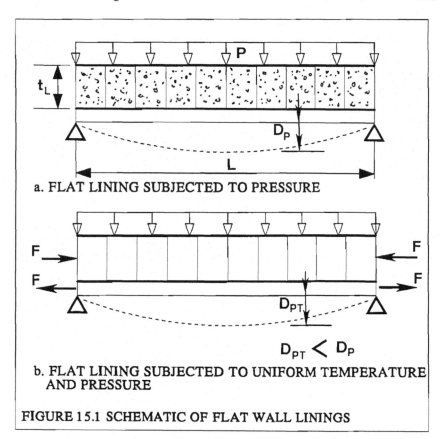

a. FLAT LINING SUBJECTED TO PRESSURE

b. FLAT LINING SUBJECTED TO UNIFORM TEMPERATURE
AND PRESSURE

FIGURE 15.1 SCHEMATIC OF FLAT WALL LININGS

lining and since no thermal restraint is imposed on the lining, the pressure is resisted solely by the support structure behind the lining. Therefore, the properties E and I are those of the support structure. Without the support structure and without any thermal displacement restraint, the lining cannot resist any pressure load.

B. Pressure Plus Uniform Temperature

The second example with regard to pressure loading on a flat brick lining includes the effects of thermal displacement restraints. In Figure 15.1b, it is assumed that the brick lining is heated to a uniform temperature T_L, while the support structure is at a lower temperature T_S. Note that no lining temperature gradient is used here. The two ends of the lining and supports are connected such that both ends of the lining and support experience the same increase in span length (ΔL). This implies that the support steel will experience a tensile loading, resulting in tensile stress across the support cross section. Likewise, the lining will experience a compressive loading across the lining cross section.

The unrestrained thermal growth ΔL_L of the lining is:

$$\Delta L_L = \alpha_L L(T_L - T_R) \tag{15.2}$$

$$= \alpha_L L \Delta T_L$$

where α_L is the lining coefficient of thermal expansion and T_R is installed lining temperature. Likewise, the thermal growth ΔL_S of the lining support is:

$$\Delta L_S = \alpha_S L(T_S - T_R) \tag{15.3}$$

$$= \alpha_S L \Delta T_S$$

where α_S is the coefficient of thermal expansion of the lining support.   Since the ends of both the lining and support are connected, the restraint displacement ΔL_R is:

$$\Delta L_R = \Delta L_L - \Delta L_S \tag{15.4}$$

The compressive displacement $\Delta L'_L$ of the lining is:

$$\Delta L'_L = F/E_L \tag{15.5}$$

where F is the resulting restraint compressive loading in the lining. An equal tensile load must be developed in the support. The tensile displacement $\Delta L'_S$ of the support is:

$$\Delta L'_S = F/E_S \tag{15.6}$$

The sum of the two displacements $\Delta L'_L$ and $\Delta L'_S$ must equal the restraint displacement, or:

$$\Delta L_R = \Delta L'_L + \Delta L'_S \qquad (15.7)$$

Substituting Equations 15.2 and 15.3 into the left side and Equations 15.5 and 15.6 into the right side, respectively, of Equation 15.7, we obtain for the expression of the restraint force F:

$$F = (\alpha_L \Delta T_L - \alpha_S \Delta T_S)[E_L E_S/(E_L + E_S)] \qquad (5.8)$$

Assuming a unit thickness of the lining (measured normal to the figure), the compressive stress (f_c) in the lining is:

$$f_{LC} = F/A_L \qquad (15.9)$$

where A_L is the lining unit cross-sectional area $(t_L \times 1)$. Similarly, the tensile stress (f_S) in the support is:

$$f_{NS} = F/A_S \qquad (15.10)$$

where A_S is the cross-sectional area of the support.

Since the lining and support are subjected to this thermal restraint loading, the pressure load displacement will now differ from the previously determined value by Equation 15.1.

Since the brick lining is exposed to a compressive loading, the brick lining can now share in resisting the pressure load. The resulting displacement due to the pressure is now expressed as:

$$D_P = 5PL^4/384(E_S I_S + E_L I_L) \qquad (15.11)$$

where E_S and I_S are for the support and E_L and I_L are for the lining. The moment of inertia of the lining (continuing with our unit depth assumption) is:

$$I_L = bt_L^3/12 = t_L^3/12 \qquad (15.12)$$

Typically, the support may be a structural shape or a combination of shapes and plate in which the I_S is found in handbooks [1].

Just as with the other lining systems, the flat brick lining joints cannot resist tensile loading. Therefore, the flat lining's capability to share in resisting the pressure loading is a function of the stress state within each joint. The limiting stress condition accounting for the full lining thickness is shown in Figure 15.2a. As the lining joints begin to separate on the back side due to the greater tensile stress, as shown in Figure 15.2b, the effective lining I_L must be revised. The lining and support each share the pressure load by the following definitions:

$$P_L = 384 \ D_p E_L I_L / 5L^4 \tag{15.13}$$

$$P_S = 384 \ D_p E_S I_S / 5L^4 \tag{15.14}$$

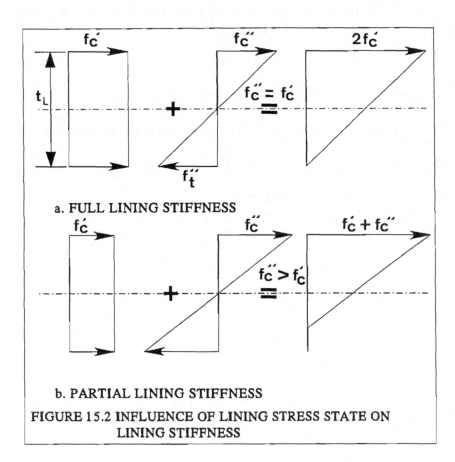

a. FULL LINING STIFFNESS

b. PARTIAL LINING STIFFNESS

FIGURE 15.2 INFLUENCE OF LINING STRESS STATE ON LINING STIFFNESS

where:

$$P = P_L + P_S \qquad (15.15)$$

and P_S is the portion of the pressure resisted by the support and P_L is the portion of the pressure resisted by the lining.

The lining bending stress (maximum at the mid-span location) is:

$$f_{LT} = 3P_L L^2 / 4 t_L^2 \qquad (15.16)$$

for the lining where f_{LT} is the mid-span lining tensile stress. The full lining thickness can be used if:

$$f_{LT} \le f_{LC} \qquad (15.17)$$

where f_{LC} is from Equation 15.9. The support bending stress is:

$$f_{BS} = PL^2 / 8S \qquad (15.18)$$

where S is the support section modulus [1]. For the support to be satisfactory, the summation of:

$$f_{ALLOW} \ge f_{NS} + f_{BS} \qquad (15.18)$$

where f_{ALLOW} is the allowable tensile stress defined in the AISE code [1], f_{NS} is from Equation 15.10 and f_{BS} is from Equation 15.18. For most structural members, f_{ALLOW} is defined as 0.60 F_y, where F_y is the support steel yield stress.

Just as with the cylindrical lining system (see Chapter 13), several trial solutions may be required to arrive at a satisfactorily effective lining thickness.

C. Temperature Gradient Through Lining

In this third case of the flat brick lining loading, a through-thickness temperature gradient is used without a process pressure loading. The lining hotface temperature T_S is defined as T_{LHF}, and the lining coldface temperature is defined as T_{LCF}. Figure 15.3a describes the lining with the through-thickness temperature gradient.

The through-thickness temperature difference causes the hotface side to expand a greater amount than the coldface side. This can be envisioned as a curling or bowing condition of the lining, as illustrated in Figure 15.3b. As shown, the lining curls toward the heat source and away from the support structure. The curling can be defined in terms of a radius of curvature R. As shown in Figure 15.3b, the thermal strain (ϵ_r) at each face of the lining is:

$$\epsilon_r = \alpha_L dT/2 \qquad (15.19)$$

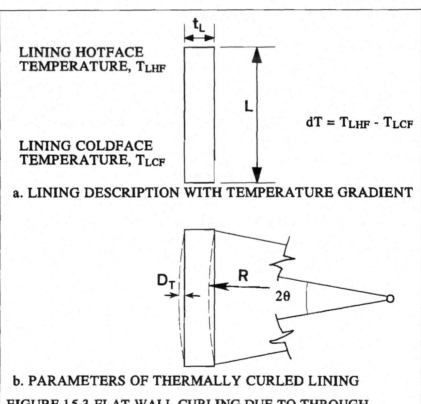

a. LINING DESCRIPTION WITH TEMPERATURE GRADIENT

b. PARAMETERS OF THERMALLY CURLED LINING

FIGURE 15.3 FLAT WALL CURLING DUE TO THROUGH-THICKNESS TEMPERATURE DIFFERENCE

For a unit length of circumference on the curved surface, the ratio of this unit length to the curved lining is identical to the ratio of the strain ϵ_r to the lining half thickness, or:

$$l/R = 2\epsilon_r/t_L$$

or:

$$R = t_L/2\epsilon_r \tag{15.20}$$

Referring to Figure 15.4, D_T is the outward displacement due to the through thickness temperature gradient. The central half angle ($\theta/2$) is defined as:

$$\theta/2 = \sin^{-1}\left(\frac{L}{2R}\right) \tag{15.21}$$

Then:

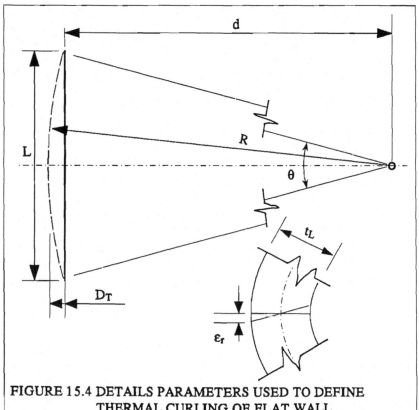

FIGURE 15.4 DETAILS PARAMETERS USED TO DEFINE THERMAL CURLING OF FLAT WALL

$$d = R\cos(\theta/2) \qquad (15.22)$$

The outward thermal displacement is:

$$D_T = R - d \qquad (15.23)$$

If the lining is not exposed to a pressure load, then the flat brick lining will curl inward and will, therefore, separate from the support.

IV. THE BEHAVIOR OF SQUARE AND RECTANGULAR REFRACTORY-LINED VESSELS

Refractory-lined vessels of circular cross-section are more stable than those of square cross section when subjected to operating loads. As shown in Figure 15.5, a comparison is made between the displacement behavior of cylindrical and square refractory-lined vessels. The vessel displacement behavior is illustrated for the normal load due to pressure (P_P) and normal load due to lining thermal expansion (P_T).

The refractory lining of the cylindrical vessel remains in full contact with the shell wall for both the operating pressure and lining thermal expansion load. Therefore, both loads develop a uniform pressure loading around the cylindrical shell.

The square vessel refractory lining, however, exhibits a significant difference in displacement behavior when subjected to operating pressure and lining expansion load. As shown in Figure 15.5b, the square vessel operating pressure load results in a uniform pressure against the shell, causing a bending displacement (and axial load) in the vessel shell. The expansion of the lining in the square vessel differs greatly from the operating pressure load.

As described in Section III.C, a through-thickness gradient will cause the lining to separate from the support structure. As illustrated in Figure 15.5b, the refractory lining thermal gradient load for the square lining causes the lining to thermally curl in and separate from the vessel shell. This would be in the absence of an

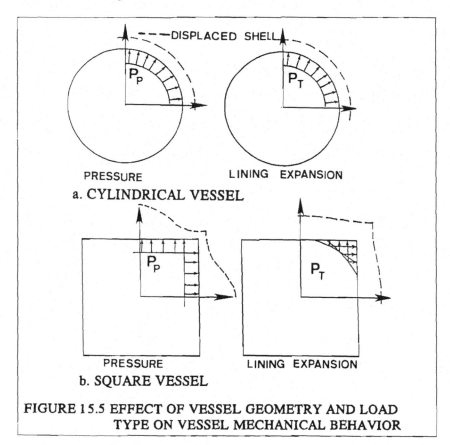

PRESSURE   LINING EXPANSION

a. CYLINDRICAL VESSEL

PRESSURE   LINING EXPANSION

b. SQUARE VESSEL

FIGURE 15.5 EFFECT OF VESSEL GEOMETRY AND LOAD
TYPE ON VESSEL MECHANICAL BEHAVIOR

operating pressure load. At the corners, however, the lining will compress into the vessel corner since the lining span (L) will typically expand more than the vessel shell. As a result, each side of the vessel lining will cause a small region of normal loading in each of the corner regions.

Example

The following example is used to estimate the magnitude of the inward lining displacement due to thermal curling that can occur in a flat wall. A square wall is 6000 x 6000 mm. The wall's hotface is 1100°C; the coldface is 500°C. The wall is assumed to be constructed of a mag-chrome brick. The wall thickness is 120 mm.

The thermal strain, using Equation 15.19, is calculated as:

$$\epsilon_r = (11.6 \times 10^{-6})(1100 - 500)/2 = 0.00348$$

The radius of curvature of the curled wall is determined using Equation 15.20:

$$R = 120/(2 \times 3.48 \times 10^{-3}) = 17,241 \text{ mm}$$

The central half-angle of the curled wall is evaluated using Equation 15.21:

$$\theta/2 = \sin^{-1}(6000/(2 \times 17,241) = 10.02°$$

The estimated magnitude of the thermal curling at the center of the wall is determined using Equations 15.22 and 15.23:

$$d = 17,241\cos(10.02/2) = 17,175 \text{ mm}$$

The amount of inward displacement due to thermal curling is:

$$D_T = 17,241 - 17175 = 65.6 \text{ mm}$$

If the wall is restrained in the plane of expansion, secondary bending can further amplify the value of D_T.

V. SUMMARY

The thermomechanical behavior of the refractory- or brick-lined square vessel differs considerably from that of the cylindrical vessel. The refractory flat lining at the mid-span regions of the square vessel is subjected to cyclic inward and outward displacements when subjected to cyclic pressure loadings. Typically, all refractory linings are exposed to through-thickness temperature gradients. With the square vessel, however, the lining displaces inward toward the heat source. The cyclic pressure loading on the square vessel will cause the lining to displace cyclically inward and outward. Therefore, the refractory at the mid-span and end regions will be subjected to cyclic stresses causing accelerated deterioration at the center and end regions of each vessel side.

A corrective measure can be taken in the design of flat

linings to eliminate or greatly reduce the tendency for the unstable and undesirable cyclic displacements. By providing an outward curvature of radius **R** and resulting offset D_T, as shown in Figure 15.6, the lining will displace outward against the lining support system when subjected to the operating temperature. The support system would have to be fabricated with this curvature to accommodate the lining curvature. Economics, in some instances, may prohibit the design of these types of lining features

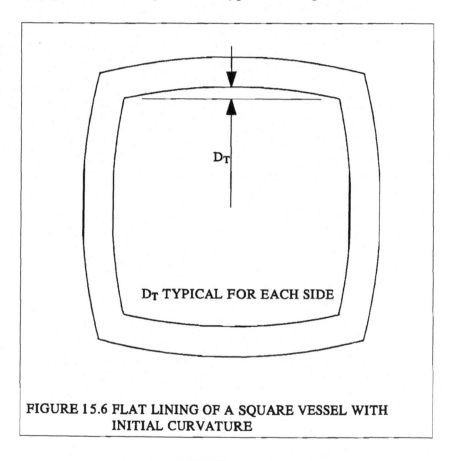

FIGURE 15.6 FLAT LINING OF A SQUARE VESSEL WITH INITIAL CURVATURE

REFERENCE

1. Manual of Steel Construction, Eighth Edition, American Institute of Steel Construction, Inc., 400 North Michigan Ave., Chicago, IL, 1980.

16
Tensile Fracture

I. INTRODUCTION

One of the most investigated modes of refractory lining failure is thermal shock fracture. Thermal shock failure is usually attributed to rapid heating or cooling [1-12] whereby tensile thermal stress develops in the refractory lining, causing localized fracture and progressive deterioration. Thermal stress fracture is recognized as a principal wear mechanism in the refractory lining of industrial process vessels. The non-linear temperature gradient during rapid heatup or cooldown develops thermal tensile stresses that fracture the refractory component with resultant lining deterioration.

The object of this chapter is to provide insight into the design of refractory materials to better resist the tensile stress environments developed in the refractory lining component under various operating conditions. Information is also provided on the mechanics of the origin of these tensile stress states and how they relate to the component fracture.

The discussion of thermal shock fracture of refractory linings is divided into two parts. Part one describes the testing of the refractory materials to evaluate their fracture strengths. Chapters 4, 5, 6, 7 and 10 addressed the concerns over refractory strength. Tensile strength of refractories has received considerable attention in the literature. Various testing procedures are used to measure the ability of the refractory material to perform under tensile stress environments [13-18]. Though testing procedures have changed over the past several decades.

The second part of a tensile fracture analysis and of our discussion consists of an investigation of a cylindrical refractory lining geometry to predict the tensile stress states in the lining system and to compare these tensile stresses against the material's susceptibility or resistance to fracture. Typically, simplified geometries have been used in the past to quantify the magnitude of thermal stresses in refractory components when subjected to transient heatup or cooldown [1]. The lining analysis described in the following sections provides a more realistic evaluation of tensile stresses in the lining. Past investigations of tensile thermal stress in lining components reverted to simplified geometries for two reasons. The analytical methods were limited to simplified closed form or algebraic equations; and the testing equipment could not be used in actual lining environments because of temperature limitations of the testing materials. That is, thermocouples and strain gages could not be exposed to the temperatures experienced by the refractories. These limitations still exist for testing equipment. Currently, however, computerized structural analysis methods allow considerable sophistication in replicating or predicting actual stress-strain conditions in various lining systems.

II. BACKGROUND

A. Material Tensile Strength

The first part in evaluating thermal fracture consists of quantifying the ultimate tensile behavior of refractory materials through material testing. The MOR test was initially the tensile strength test to estimate the refractory material ultimate tensile strength. As discussed in Chapter 4, the MOR test is typically a three-point bending test in which a load is increased until fracture occurs. This test is a stress-controlled load test (see Chapter 2) and does not measure the ultimate tensile strain. Since thermal tensile stress is a strain-controlled load, the MOR test is not applicable and does not provide insight into thermal tensile fracture. That is, the ultimate strain is not known for comparison with the thermal tensile strain. If the tensile stress-strain data are known, the ultimate strain can be estimated from the MOR test (assuming the compressive stress-strain data are also known). Likewise, from analytical procedures, the stress-strain data allow a definition of the thermal stress when the thermal strain conditions are known.

The work-of-fracture (WOF) test is a modification of the MOR test, in which a notch geometry has been added to the MOR test specimen [14-15]. More importantly, the displacement (or strain) is also measured simultaneously with the applied loading. Basically, the load-displacement material relationship is incorporated into the test which couples the refractory stress-strain behavior with crack geometry, making the work-of-fracture test more applicable in measuring a material's strength in resisting thermal tensile stress. However, the objectives in refractory material design to achieve greater WOF values is perhaps misunderstood.

The work-of-fracture tests lead to the application of a fracture mechanical approach to refractories [13]. The stress intensity K_I at a crack tip opening mode due to an applied stress σ_{app} for a crack length C is defined as:

$$K_I = \sigma_{app} Y C^{0.5} \qquad (16.1)$$

where Y is a crack geometry factor. At fracture, Equation 16.1 becomes:

$$K_{IC} = \sigma_f Y C_{crit}^{0.5} \qquad (16.2)$$

where K_{IC} is known as the fracture toughness, σ_f is the tensile fracture stress and C_{crit} is the critical crack length. Equation 16.3 [13] provides a relationship between Equations 16.1 and 16.2:

$$V = A(K_I / K_{IC})^N \qquad (16.3)$$

where V is the crack velocity (rate of crack growth), A is a constant for the material crack geometry and N is a parameter describing the slow crack growth resistance of the subject refractory lining material. Larger N values indicate larger crack resistance of the refractory lining material.

Material tests are conducted to evaluate the parameters in Equations 16.1, 16.2 and 16.3. Equation 16.2 identifies σ_f as the material property to increase for greater fracture toughness of the material. If the material is subjected primarily to stress-controlled loads, then optimization of σ_f is appropriate. However, for strain-controlled loading Equation 16.2 should be expressed as:

$$K_{IC} = \epsilon_f \, YC^{0.5}E \qquad (16.4)$$

With the objective of increasing ϵ_f while decreasing E in designing refractory materials to resist transient and cyclic thermal strains.

The third area of investigative work on the fracturing of refractory materials is with respect to crack growth resistance, R. The crack growth resistance is expressed as [19-21]:

$$R = \sigma_f \, (1 - \nu)/\alpha E \qquad (16.5)$$

where ν is Poisson's ratio and α is the coefficient of thermal expansion. The material crack growth resistance is usually thought to be improved by increasing the σ_f of the material. When this is done, the material's E will most often also increase. Basically, the crack growth resistance equation, when expressed in the form of thermal strain, can be defined as:

$$R = \epsilon_f \, (1-v)/\alpha \qquad (16.6)$$

Where ϵ_f is the ultimate strain at ultimate strength. The other designation of crack growth resistance are expressed as:

$$R' = \sigma_f \, (1-v)k \, /\alpha E \qquad (16.7)$$

And:

$$R'' = \sigma_f \, (1-v)a/\alpha E \qquad (16.8)$$

where k is the thermal conductivity and a is the thermal diffusivity. Since these equations are addressing thermal strains, or strain-controlled loading, it appears that it would be more appropriate to express Equations 16.7 and 16.8 also as:

$$R' = \epsilon_f \, (1-v)k \, /\alpha \qquad (16.9)$$

And:

$$R'' = \epsilon_f \, (1-v)a/\alpha \qquad (16.10)$$

Since the crack growth resistance is typically addressed to thermal loading, the testing done to evaluate the various R values should be conducted using strain controlled loading. Also, the desirable refractory material property to increase would be ϵ_t and not S_t.

Figure 16.1 shows the refractory material's desired work-of-fracture curves for both stress-controlled and strain controlled loads.

B. Tensile Stress State in Restrained Lining

Refractory material has historically been treated as a homogeneous isotropic linear elastic material and the lining component has been treated as a simple geometry, isolated and free

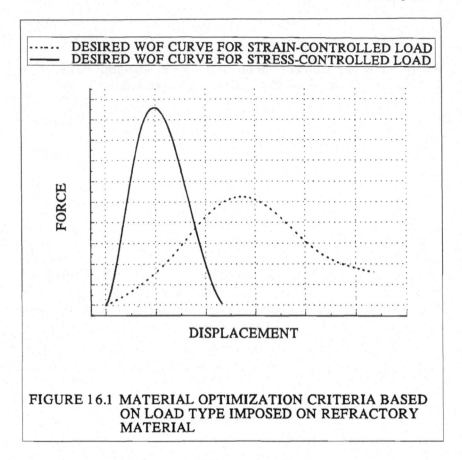

FIGURE 16.1 MATERIAL OPTIMIZATION CRITERIA BASED
ON LOAD TYPE IMPOSED ON REFRACTORY
MATERIAL

of external restraint. Although recent studies [22-24] on thermal stress fracture continue to treat the lining refractory as an isolated traction free component, a more realistic approach has been used to deal with the stress-strain behavior of the component as a two-dimensional component shape.

Even though a major portion of refractory linings used in industrial applications are contained in steel vessel shells that restrain thermal growth of the lining, little attention has been given to the influence of the shell restraint on thermal fracture. It has been shown [25-33] that shell restraint greatly alters the thermomechanical behavior of the refractory lining and adds complexity in predicting thermal stress fracture during rapid heating and cooling. The objective of this section is to provide an understanding of the influence of lining restraint, combined with the influence of non-linear elastic/plastic behavior of refractory materials on thermal stress fracture and resulting lining deterioration.

The study is limited to a cylindrical lined vessel subjected to typical transient heatup, steady-state and cooldown conditions. Finite element analysis is used to identify the locations of maximum tensile stresses in the lining component as well as the direction of maximum tensile stress.

Thermal fractures that occur a short distance from the lining hotface and run parallel to the hotface are frequently referred to as *slabbing, spalling, flaking* or *peeling.* A second form of thermal fracture originates at the mid-region of the hotface and appears to propagate perpendicular to the hotface toward the interior of the brick component. The literature does not define name for this mode of fracture. *Pinch spalling* is a third form of lining deterioration and is a crushing action at the hotface corner region of the radial brick joints. Pinch spalling is usually associated with concentrated compressive stresses which cause shear fracture of the brick. The latter stages of pinch spalling are referred to as *cobbling.*

As a result, there are four primary concerns addressed in the following study. They are:

1.　　Does the predicted location and direction of the various maximum thermal tensile stress obtained from the restrained lining analysis assist in improving our understanding of the various modes of thermal tensile fracture occurring within the lining component?

2.　　Does the vessel shell restraint greatly influence the stress state that occurs within the lining component?

3.　　Does the mix of vessel shell restraint and thermal conditions (heatup, cooldown and steady-state) cause tensile stress states in the lining component that provide insight into which combinations are the major contributors to the cause of fracturing of the lining component?

4.   Does the plastic deformation which should be greatest at the lining hotface assist in causing fracturing of the lining hotface?

5.   Does the tensile stress state in a wide brick component differ from that of a narrow brick component?

III. CYLINDRICAL REFRACTORY-LINED VESSEL ANALYSIS: BACKGROUND

A. Introduction

The cylindrical refractory-lined vessel used for the analysis of the restrained lining is described in Figure 16.2. As previously discussed, the objective of this analysis is to determine the impact of the vessel shell restraint on the tensile stress state in the lining during typical operating temperatures.

The following conditions were used for the vessel design. The vessel shell was typical carbon steel. The lining is a high-alumina (90%) fired brick.

A summary of the lining and shell geometry and material properties are summarized in Table I. As shown, the total lining thickness is set at 252 mm for the component I lining design and 305 mm for the component II lining design. Two extreme sizes of the lining component (brick shape) were used in this investigation. Lining No. I brick component (see Figures 16.3 and 16.4) is 305 mm wide in the circumferential direction and is 152 mm in radial thickness. Figure 16.3 illustrates the unrestrained lining components and Figure 16.4, the restrained lining components. A backup lining of 102-mm thickness was used, making a total lining thickness of 305 mm. For the purposes of this investigation, the

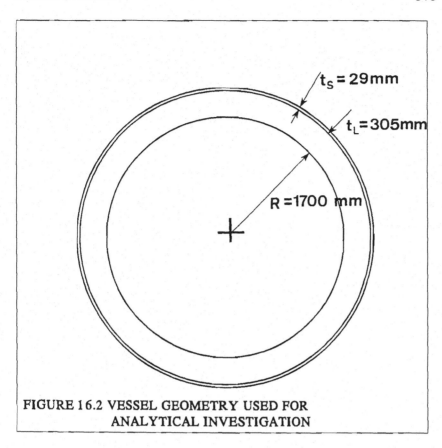

FIGURE 16.2 VESSEL GEOMETRY USED FOR
ANALYTICAL INVESTIGATION

backup lining and the work lining joints were not staggered. The dimensions for the second lining (component II, Figures 16.3 and 16.4) are the reverse from No. I. The No. II component has a 152-mm circumferential width and a 305-mm radial thickness. Therefore, the total lining thickness consists of one component with the coldface against the shell.

The two component geometries were selected because the work by Bradley et al. [22-24] showed that the width (W)-to-thickness (T) ratio of the work lining component had an influence on the magnitude and location of the maximum tensile stress that would cause fracture. It should be emphasized, however, that Bradley's work did not include the restraint of the vessel shell.

Table I

Parameter Values Used in
Investigative Analysis

Component	Parameter	Value
Steel Shell	Elastic Modulus (E_s)	207 GPa
	Poisson's Ratio (v_s)	0.3
	Thermal Expansion Coefficient (α_s)	11.7×10^{-6} °C^{-1}
	Thermal Diffusivity (a_s)	0.11 cm^2 s^{-1}
	Thermal Conductivity (K_s)	36 Wm^{-1}K^{-1}
	Shell Thickness (t_s)	29 mm
No. I & II Brick Component	Elastic Modulus (E_B)	See Table III
	Poisson's Ratio (v_B)	0.3
	Thermal Expansion Coefficient (α_B)	10×10^{-6} °C^{-1}
	Thermal Diffusivity (a_B)	0.01 cm^2 s^{-1}
	Thermal Conductivity (K_B)	2.07 Wm^{-1}K^{-1}
No. I*	Length (L) Width (W)	152 mm 302 mm
No. II	Length (L) Width (W)	302 mm 152 mm

*For lining component between No. I component and shell,
 W = 302 mm, L = 102 mm.

B. Analysis Methodology

The cylindrical refractory-lined vessel was analyzed using the finite element method [34]. The model of the unrestrained and restrained lining systems considered are described, respectively, in Figures 16.5 and 16.6. Because of symmetry of geometry and

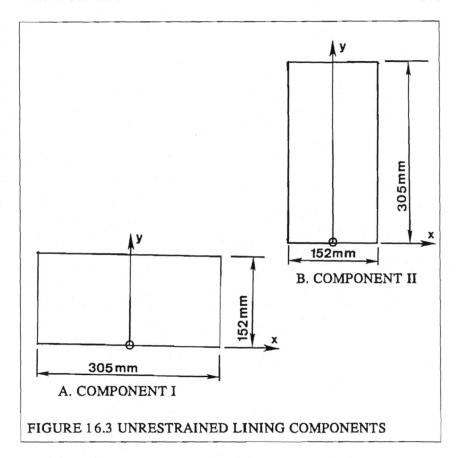

FIGURE 16.3 UNRESTRAINED LINING COMPONENTS

thermal loading, only half of the circumferential width of each of the two lining systems was required to be included. Figure 16.6 identifies the joints, each lining component geometry and the shell geometry. The joints were modeled with a non-linear element that would transfer a compressive load across the joint and simulate the joint in a contact mode. If a tensile load were applied to a joint, the joint would separate. As discussed previously, the primary objective of this investigation is to evaluate the influence of vessel shell restraint on the stresses developed in the lining components. However, for the purposes of comparison, the same components were also analyzed without the lining restraint.

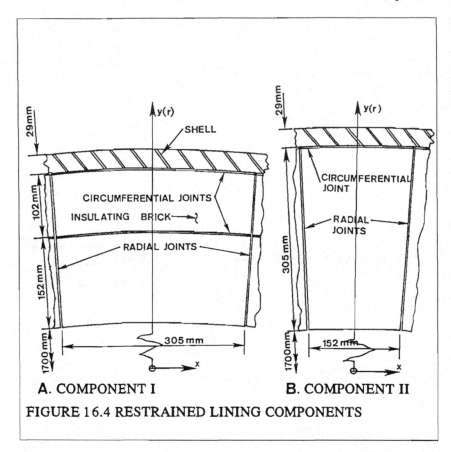

A. COMPONENT I   **B.** COMPONENT II

FIGURE 16.4 RESTRAINED LINING COMPONENTS

The vessel lining stresses were evaluated for a heatup transient, two cooldown transients (see Figure 16.7 and Table II)-- each starting from two different steady-state lining temperatures-- and two steady-state lining temperatures. Figure 16.8 describes the through-thickness temperature profiles of the five thermal conditions. The two cooldown transients start from the two described steady-state lining temperatures. The transient heatup and cooldown rates were both set at 10°C per minute. This transient rate was selected as being representative of some industrial rapid heatup practices and is of a sufficient rate to cause transient thermal stresses. The lining thermal stresses were evaluated when the heat-up transient caused the lining hotface to reach about 630°C. The lining thermal stresses were evaluated during the first cooldown

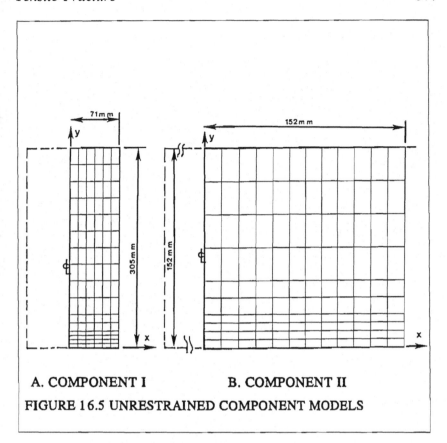

A. COMPONENT I   B. COMPONENT II

FIGURE 16.5 UNRESTRAINED COMPONENT MODELS

when the lining hotface reached about 500°C. Note that this cooldown is from the higher steady-state lining temperature condition. This first cooldown starts from a lining hotface temperature of 1100°C. The lining thermal stresses were also evaluated for the second cooldown transient when the lining hotface reaches 50°C. This second cooldown starts from a lining steady-state temperature condition in which the lining hotface is 650°C. The lining stresses are evaluated for the heatup transient when the lining hotface reached about 650°C. Lining component I's total lining thickness was less than lining component II's total lining thickness. However, the transients did not have any significant influence on either of the linings' coldface or shell temperatures. As shown in Figure 16.8, the 152-mm portion of lining measured

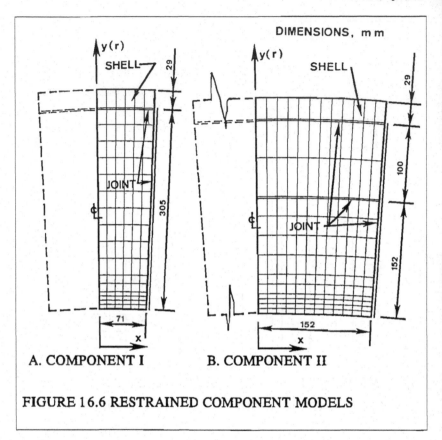

A. COMPONENT I   B. COMPONENT II

FIGURE 16.6 RESTRAINED COMPONENT MODELS

from the hotface represents the extent of penetration of the heatup and cooldown transients. Therefore, the 152-mm portions of the cooldown and steady-state temperature profiles were applied to the No. I component. The transient through-thickness temperature profiles, T(y), were calculated from the classical analytical solution [35] for a semi-infinite slab over a depth range $0 < y < L$:

$$T(y) = 4 \, \phi \, ti^2 \, erfc \left[\frac{y}{2\sqrt{at}} \right]$$

In the case of the restrained condition of components No. I and No. II, the steel shell was assumed to have no through-thickness temper-

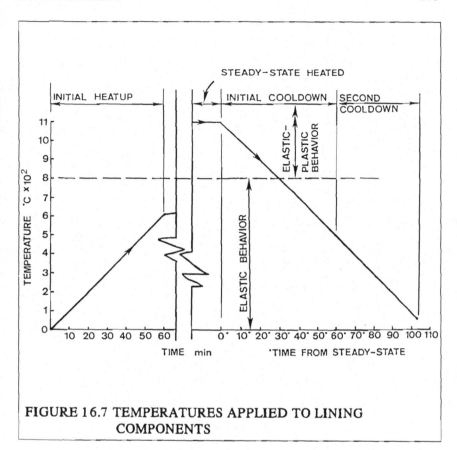

FIGURE 16.7 TEMPERATURES APPLIED TO LINING
                COMPONENTS

ature gradient and the shell temperature was made identical to the coldface temperature of the lining. A classical analytical solution was used to define the through-thickness transient temperature distribution in the unrestrained and restrained lining systems. A finite element thermal analysis would have produced the same results. The closeup form solution was used simply for ease, in this particular case, for defining the lining temperatures.

In the analytical investigation of thermal stresses, identical temperature profiles and thermophysical material properties were used for both restrained brick component geometries. A hypothetical elastic-plastic stress-strain behavior was assumed for the refractory material. Bilinear stress-strain curves were used to

Table II

Through-Thickness Temperature Profiles
Used for Evaluating Thermal Stress in No. I and
No. II Lining Components: Unrestrained
and Restrained

Temperature Profile No.	Description of Temperature Condition
1	Heatup at 10°C/min for one hour from a uniform lining ambient temperature of 20°C
2	Steady-state lining temperature with a 650°C hotface and 180°C shell temperature
3	Steady-state lining temperature with a 1100°C hotface and 260°C shell temperature
4	Cooldown at -10°C/min. for one hour from a steady-state lining temperature with a 1100°C hotface, 260°C shell temperature
5	Cooldown at -10°C/min. for one hour from a steady-state lining temperature with a 650°C hotface and 180°C shell temperature

represent the non-linear stress-strain curves and to simulate the elastic and plastic behavior of the refractory material. As illustrated in Figure 16.9, and described in Table III, three parameters were used to represent each temperature-dependent bilinear stress-strain curve: elastic modulus (E_e), plastic modulus (E_p) and yield point (S_y). As shown, the material properties are highly temperature-dependent (Figure 16.10). It should be noted that the threshold of plastic deformation in the refractory material was approximately 800°C. Therefore, plastic deformation of the refractory lining material will increase as the lining temperature increases above 800°C.

Important differences exist between the stress-strain behavior

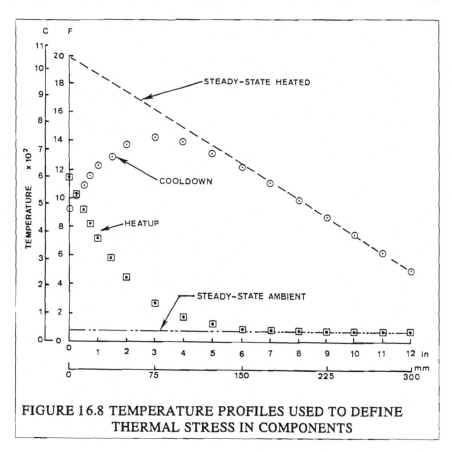

FIGURE 16.8 TEMPERATURE PROFILES USED TO DEFINE
THERMAL STRESS IN COMPONENTS

of a restrained and unrestrained lining component. The fundamental
parameters that are used in the analysis of a rapidly heated
unrestrained linear elastic refractory lining component are illustrated
in Figure 16.11. For the unrestrained lining component, the thermal
stress developed within the component is a function of d^2T/dy^2. In
the case of a linear temperature gradient in an unrestrained
component, $d^2T/dy^2 = 0$, and no thermal stress exists. It will be
shown that the linear through-thickness temperature distribution in a
restrained lining causes significant lining thermal tensile stresses.

In the steel vessel shell, thermal growth of the refractory
lining is restrained causing circumferential compressive loading of

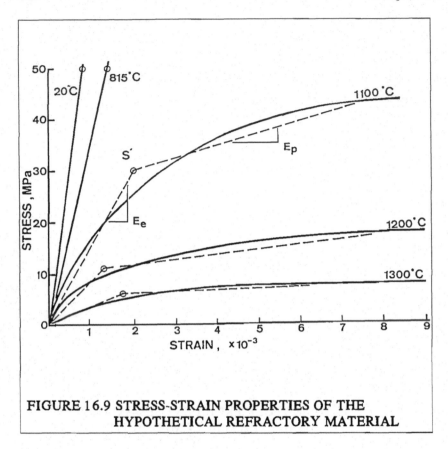

FIGURE 16.9 STRESS-STRAIN PROPERTIES OF THE
HYPOTHETICAL REFRACTORY MATERIAL

the radial joint as illustrated in Figure 16.12. With initial lining
heatup, a circumferential compressive restraint load occurs over a
partial length of the radial joint (see Chapter 13). As the
temperature increases, the restraint load increases and is distributed
over a greater portion of the radial joint. Assuming elastic lining
component behavior, as illustrated in Figure 16.13, the compressive
restraint load will decrease from a maximum value at the hotface of
the radial joint to a negligible value at the interior location of the
radial joint. Because of the assumption regarding joint behavior
that no tensile stress can be developed in the radial (and
circumferential joints), the remainder of the joint opens.

During cooldown of a restrained heated lining, thermal lag
causes faster contraction on the hotface portion of the elastic lining

Table III

Mechanical Properties of Hypothetical Refractory Lining Material

Tempera-ture °C	Elastic Modulus (E_e) GPa	Plastic Modulus (E_p) GPa	Yield Stress S_Y MPa	Poisson's Ratio	Density kg/m^3
20	60	58.8	50	0.2	2490
800	33	32.3	50	0.2	2490
1100	15	3.6	30	0.2	2490
1200	8.6	1.3	11.5	0.2	2490
1300	3.6	0.3	6.2	0.2	2490

component than on the interior coldface portion of the component. Therefore, during cooldown the hotface end of the radial joint will tend to open, resulting in a radial joint restraint load, as described in Figure 16.14.

A maximum radial tensile stress S_Y (stress in Y direction parallel to joint, see Figure 16.15) can occur along the radial joint in the restrained lining component during rapid heating or cooling as well as during the steady-state heated condition. The partially compressed radial joint causes a maximum radial tensile stress at the tail of the restraint load. The mathematical explanation for the maximum radial tensile stress at the tail of the restraint load is illustrated in Figure 16.10. For the steady-state linear temperature profile without a lining restraint, no thermal stress is developed in the lining component:

$$d^2T/dy^2 = f\left[\alpha d^2T/dy^2\right]$$
$$= f\left[d^2\epsilon_U/dy^2\right] = 0$$

where T is the temperature at location y, α is the coefficient of thermal expansion and ϵ_U is the unrestrained thermal expansion at location y.

However, with the restraint load applied over a partial length of the radial joint, a change occurs in the slope of the strain be-

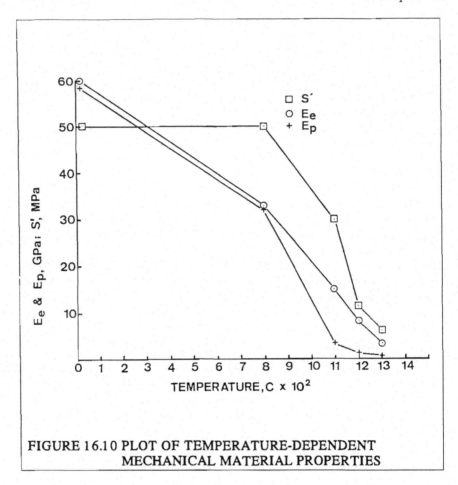

FIGURE 16.10 PLOT OF TEMPERATURE-DEPENDENT
MECHANICAL MATERIAL PROPERTIES

tween the unrestrained and restrained portion of the refractory component. At the tail of the restraint load, a change in the slope of strain occurs, resulting in a discontinuity of the strain:

$$d^2\epsilon/dy^2 > 0$$

At this point of discontinuity, a maximum radial tensile stress occurs, as illustrated in Figure 16.12. Therefore, in a restrained lining a maximum radial tensile stress will develop at the tail of the restraint load along the radial joint, regardless of whether the lining is exposed to transient or steady-state temperature conditions.

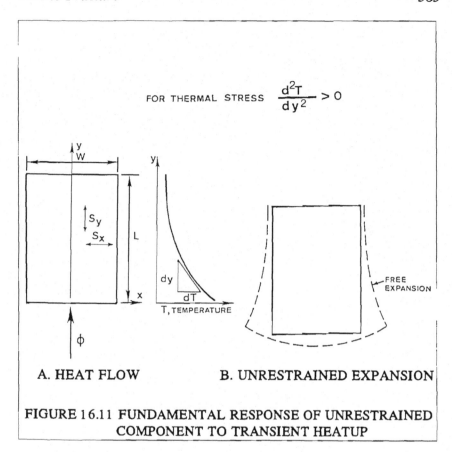

FOR THERMAL STRESS $\dfrac{d^2T}{dy^2} > 0$

A. HEAT FLOW   B. UNRESTRAINED EXPANSION

FIGURE 16.11 FUNDAMENTAL RESPONSE OF UNRESTRAINED
COMPONENT TO TRANSIENT HEATUP

During cooling of a restrained heated lining, a tail is developed at each end of the restraint load, resulting in two local maximum radial tensile stresses along the radial joint.

IV. RESULTS OF CYLINDRICAL REFRACTORY-LINED VESSEL ANALYSIS

A. Introduction

The results of the thermal stress analysis are described in two parts. Part one consists of a discussion about the tensile thermal stresses developed at the end of the heatup (profile 1) and during the two steady-state thermal conditions (profiles 2 and 3). There is

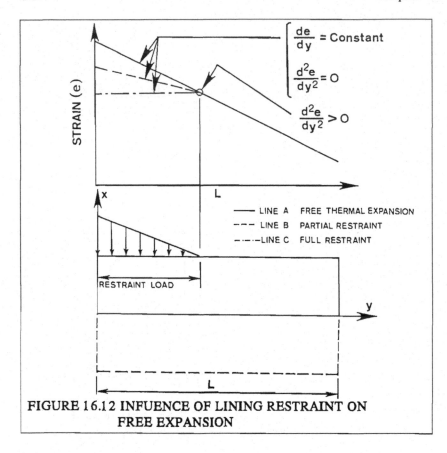

FIGURE 16.12 INFUENCE OF LINING RESTRAINT ON FREE EXPANSION

a question as to which creates the maximum tensile fracture stress--the transient heatup tensile stress or the steady-state tensile stress.

The second part of the discussion is on the two cooldown transients (profiles 4 and 5). The profile 4 cooldown starts from a higher steady-state temperature. Therefore, the hotface region of components I and II will have been exposed to plastic deformations. The profile 5 cooldown is from a much lower steady-state temperature at which plastic deformations do not occur. Therefore, the coupling of plastic tensile stress and transient tensile stress will not occur in the profile 5 cooldown analysis.

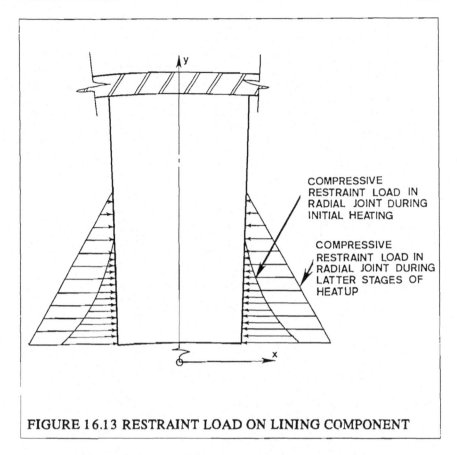

COMPRESSIVE
RESTRAINT LOAD IN
RADIAL JOINT DURING
INITIAL HEATING

COMPRESSIVE
RESTRAINT LOAD IN
RADIAL JOINT DURING
LATTER STAGES OF
HEATUP

FIGURE 16.13 RESTRAINT LOAD ON LINING COMPONENT

B. Component I Stress Analysis Results for Heatup and Steady-State Conditions (Profiles 1, 2 and 3)

The results of the thermal stress investigation of the restrained and restrained component I [26,27] are summarized by describing the stress behavior. Since the maximum temperature at the lining hotface is 650°C for profiles 1 and 2, no inelastic straining occurred. That is, the component remained elastic. For profile 3 steady-state temperature, the 1100°C hotface caused plastic straining.

The plot of the X-direction stress, S_x, [26] of the unrestrained component I transient heatup is described in Figure 16.16 [26]. The maximum tensile S_x stress (stress component parallel to hotface)

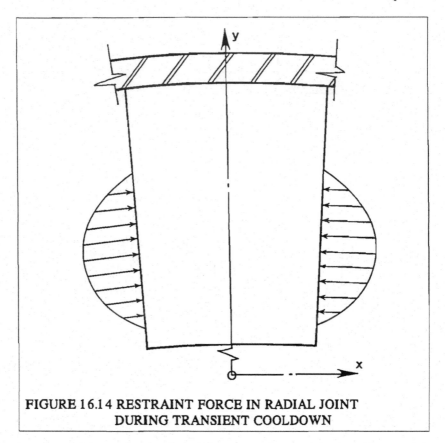

**FIGURE 16.14 RESTRAINT FORCE IN RADIAL JOINT
DURING TRANSIENT COOLDOWN**

is maximal (+37.63 MPa) at the interior center region of this lining component. If the stress were sufficient in magnitude to cause fracture, the fracture due to the S_X stress would initiate at this interior location in the vertical direction parallel to the hotface.

The plot of the Y-direction stress, S_Y, of the unrestrained component I transient heatup is described in Figure 16.17. The maximum tensile S_Y stress (7.87 MPa) occurs at an interior location near the face edge of component I. Comparing the magnitudes of S_X and S_Y the tensile fracture would initiate first at the maximum S_X location.

The plot of the shear stress, S_{XY}, of the unrestrained component I transient heatup is described in Figure 16.18. The

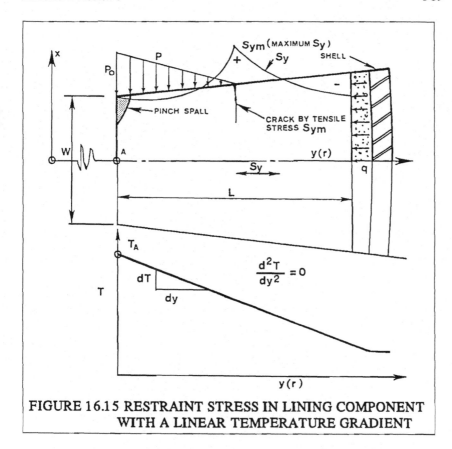

**FIGURE 16.15 RESTRAINT STRESS IN LINING COMPONENT
WITH A LINEAR TEMPERATURE GRADIENT**

maximum shear stress (9.5 MPa) occurs near the hotface corner.

The plot of the X-direction stress of the restrained component I transient heatup is shown in Figure 16.19 [27]. The restraint of the vessel shell has a significant influence on the stress state within component I in that no tensile S_x stresses exist. The maximal compressive stress (284 MPa) is at the hotface. This is a rather high compressive stress and is most likely due to the artificially high MOE used for the refractory lining material. For lesser MOE, all of the stresses would be scaled accordingly. However, the results would be similar with respect to location of maximum stress values.

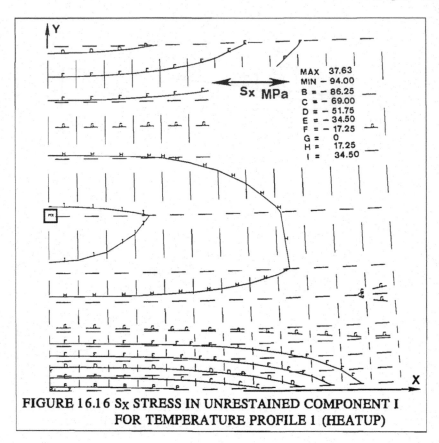

FIGURE 16.16 Sx STRESS IN UNRESTAINED COMPONENT I
FOR TEMPERATURE PROFILE 1 (HEATUP)

The plot of the Y-direction stress of the restrained component I transient heatup is described in Figure 16.20. The maximal tensile S_Y stress is 32 MPa and is located at the end region of the constraint force. This condition was previously described in Figure 16.12. These results indicate that the slabbing crack would initiate at the interior end portion of the radial joint that is in compression. This crack would initiate parallel to the hotface and propagate inward toward the center of the component. According to the scale of the loading, this is about 50 mm in from the hotface.

The shear stress plot of the restrained component I transient heatup is shown in Figure 16.21. The maximum is 28.1 MPa and occurs at the hotface end of the radial joint, the location of the joint

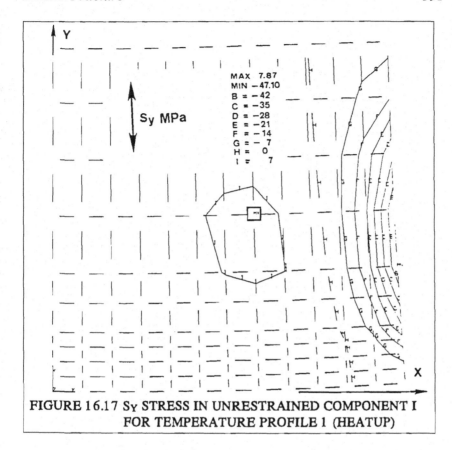

FIGURE 16.17 Sy STRESS IN UNRESTRAINED COMPONENT I
FOR TEMPERATURE PROFILE 1 (HEATUP)

with the highest S_x compressive stress (see Figure 16.19). This is the region of the brick joint in which pinch spalling is normally observed.

The results of the component I transient heatup stresses for both the unrestrained and restrained conditions are summarized in Figure 16.22. The circles identify the location of the maximum stress, and the line through the circle identifies the crack orientation. Note that the restrained condition was evaluated using both elastic and elastic-plastic material behavior. Although the hoop stress, S_x, differed, imperceptible differences were observed in the S_Y tensile stress at the radial joint. The steady-state heated condition (hotface of 650°F) are also included in Figure 16.22.

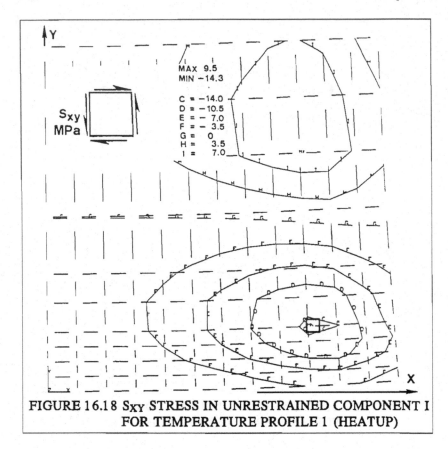

FIGURE 16.18 S$_{XY}$ STRESS IN UNRESTRAINED COMPONENT I
FOR TEMPERATURE PROFILE 1 (HEATUP)

Since the linear through-thickness temperature gradient causes
no tensile thermal stress in the unrestrained component I, that
portion is omitted from Figure 16.22. For profiles 2 and 3 steady-
state temperature gradients on the restrained component I, the full
length of the radial joint is in compression and, therefore, no S$_Y$
tensile stress component exists at the lower steady-state temperature
condition.

It can be concluded that for the heatup condition, the
component I slabbing spall is due to the vessel shell restraint and
the compressive loading over the partial length of the radial joint.
At steady-state temperature, the joint is in full compression and no
tensile stress exists to cause slabbing. The slabbing tensile stresses

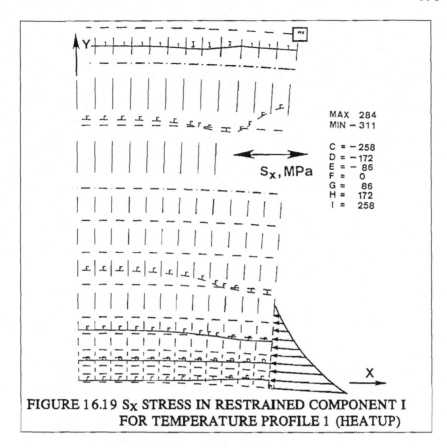

FIGURE 16.19 S_X STRESS IN RESTRAINED COMPONENT I FOR TEMPERATURE PROFILE 1 (HEATUP)

due to restraint are considerably greater than the slabbing tensile stresses that are developed for the unrestrained condition.

Figure 16.23 describes the amount of elastic and plastic straining that occurs in component I due to the combination of the higher temperature of profile 3 and the vessel shell restraint. The elastic and plastic strain are about equal in magnitude for the conditions assumed. The resulting distribution of the hoop stress, or S_X stress, is described in Figure 16.24. Note that the hotface compressive S_X stress is less than the interior S_X compressive stress as a result of the plastic straining at the hotface. The higher thermal strain and higher temperature at the hotface region result in the greater plastic straining at the hotface region.

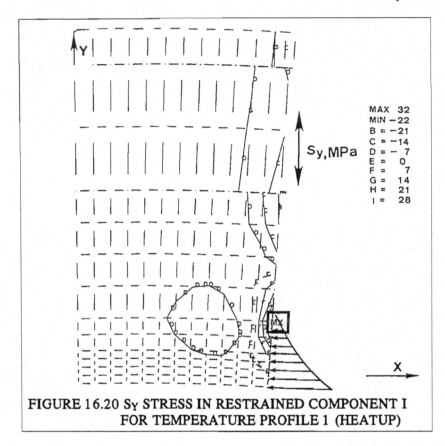

**FIGURE 16.20 S_Y STRESS IN RESTRAINED COMPONENT I
FOR TEMPERATURE PROFILE 1 (HEATUP)**

C. Component II Stress Analysis Results for Heatup and
Two Steady-State Conditions (Profiles 1, 2 and 3)

The results of the thermal stress investigation of the unrestrained and restrained component II [26, 27], subjected to temperature profiles 1, 2 and 3, are similar in behavior to the previous results shown for the component I investigation with the exception of the steady-state profile results.

The two component II steady-state temperature profiles caused the restraint load to vary in magnitude and caused the joint to compress over different lengths, as shown in Figures 16.25 and 16.26. The component II, unlike the component I, did not have

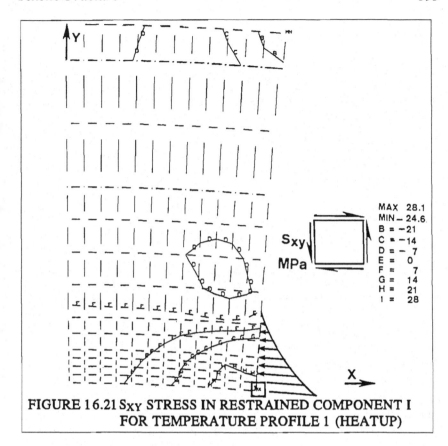

FIGURE 16.21 S_{XY} STRESS IN RESTRAINED COMPONENT I
FOR TEMPERATURE PROFILE 1 (HEATUP)

a circumferential joint. Therefore, the longer continuous radial joint surface of component II resulted in S_Y tensile stresses as shown. The maximal tensile S_Y stress at the radial joint for the profile 2 steady-state temperature was about 16 MPa while the greater profile 3 steady-state temperature caused a maximum S_Y of 26 MPa.

The reason for the component I having no S_Y radial joint tensile stresses is explained in Figures 16.27 and 16.28. The compressive restraint force causes component I to bend. However, the circumferential joint opens, resulting in no radial joint S_Y tensile stress.

A summary of the S_X and S_Y maximum tensile stresses is

RESTRAINT CONDITION	ANALYSIS TYPE	HEATUP TRANSIENT		STEADY-STATE HEATED	
		S_x, MPa	S_y	S_x	S_y
RESTRAINED		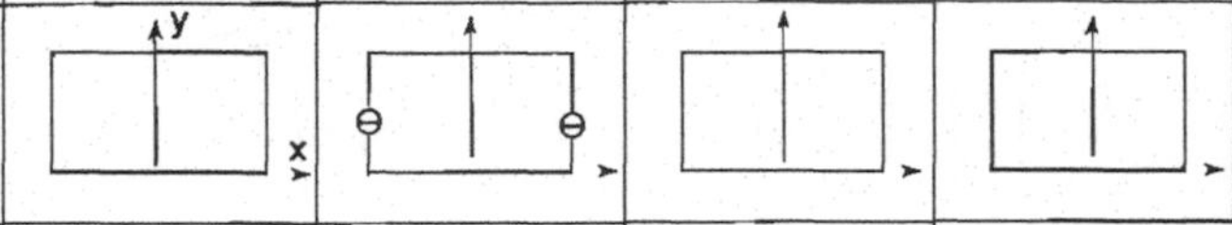			
	ELASTIC	0	32	0	0
	ELASTIC-PLASTIC	0	32	0	0
UNRESTRAINED		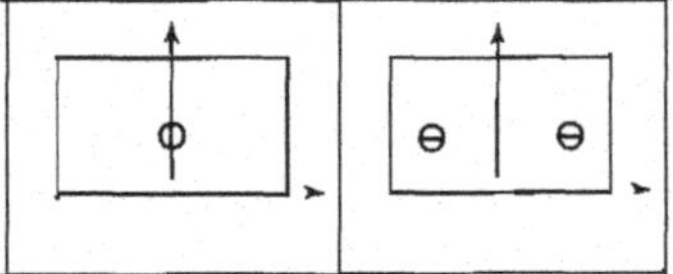			
	ELASTIC	38	8		

FIGURE 16.22 SUMMARY OF MAXIMUM THERMAL TENSILE STRESS IN COMPONENT I FOR HEATUP AND STEADY-STATE PROFILES 1, 2 AND 3

described in Figure 16.29. As with component I, the elastic versus elastic-plastic refractory lining behavior did not greatly affect the magnitude of predicted maximum tensile stresses. The steady-state condition stresses described in Figure 16.29 are for profile 3, with the hotface temperature of 1100°C. The maximum tensile stresses during heatup occur at the tail of the constraint load in the radial joint for the restrained component II and are similar in magnitude to the unrestrained component II tensile stresses that occur at interior region.

The maximum tensile stress for the steady-state temperature (profile 2, hotface at 650°C) of the restrained component II is about 6 MPa (see Figure 16.30). The profile 2 tensile stresses are similar in magnitude to the end of the heatup tensile stresses.

D. Component I Stress Analysis Results for the Two Cooldown Transients (Profiles 4 and 5)

The first cooldown results for profile 4 in which the cooldown starts from the hotface steady-state temperature of 1100°C. The thermal stresses are evaluated at one hour of cooldown from the steady-state profile 3. The S_x and S_y tensile

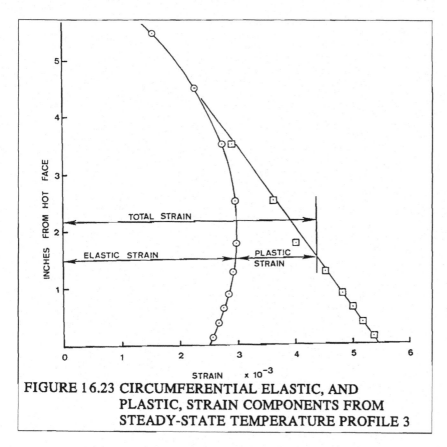

FIGURE 16.23 CIRCUMFERENTIAL ELASTIC, AND
PLASTIC, STRAIN COMPONENTS FROM
STEADY-STATE TEMPERATURE PROFILE 3

thermal stresses for the elastic-plastic analysis is described in Figures 16.30 and 16.31, respectively. As shown, a short length of the radial joint, in the region of the hotface, is open. During cooldown, two maximum thermal tensile stress states are developed. One at the tail of the radial joint compressive load and the second at the center point of the hotface. The results are summarized in Figure 16.32. Note that the elastic-plastic analysis results for the profile 4 cooldown shows higher thermal stresses than those obtained from the elastic analysis. The higher stresses from the elastic-plastic analysis show that the plastic straining that occurs at the profile 3 steady-state temperatures has a significant influence in amplifying the cooldown thermal stress. The elastic analysis results are only showing the effects of transient cooldown

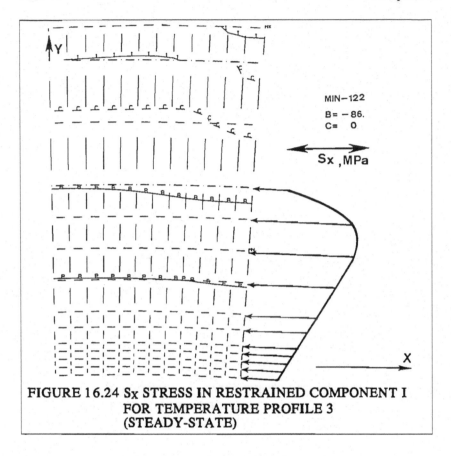

FIGURE 16.24 S_X STRESS IN RESTRAINED COMPONENT I
FOR TEMPERATURE PROFILE 3
(STEADY-STATE)

profiles. The compressive restraint load assists in reducing the transient cooldown tensile stresses.

The second cooldown (profile 5) was from a lower steady-state temperature (profile 2) at which a considerably less amount of plastic straining occurred. Also, less compressive restraint loading was imposed on the lining at these lower temperatures. The result was that the cooldown from the lower steady-state condition resulted in higher tensile stresses, as described in Figure 16.32. Note the slightly lesser tensile stresses in the unrestrained component I. The location and direction of tensile cracking are similar for both the restrained and unrestrained component I.

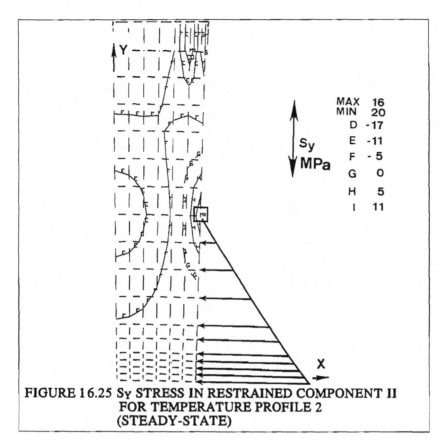

MAX 16
MIN 20
D -17
E -11
F - 5
G 0
H 5
I 11

S_Y MPa

**FIGURE 16.25 S_Y STRESS IN RESTRAINED COMPONENT II
FOR TEMPERATURE PROFILE 2
(STEADY-STATE)**

E. Component II Stress Analysis Results for the Two Cooldown Transients (Profiles 4 and 5)

The component II cooldown (profile 4) tensile thermal stresses are described in Figure 16.33. These results are for the elastic-plastic refractory material behavior. As with component I, the cooldown stresses are evaluated after one hour of cooldown. The radial joint opens in the region of the hotface. As a result, two maximum S_Y tensile stress conditions exist along the radial joint. The maximum tensile S_Y stress (46 MPa) occurs at the end of the open joint in the hotface region. The second maximum tensile stress (30 MPa) occurs on the back side of the radial joint at the second end of the open joint.

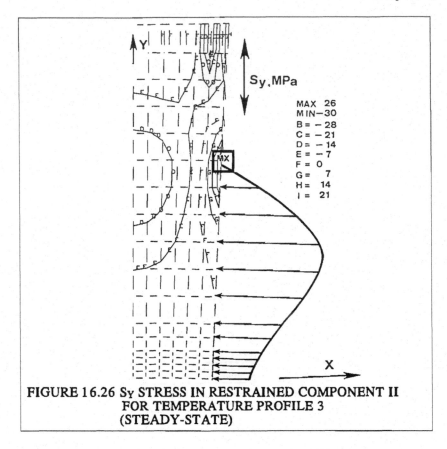

**FIGURE 16.26 Sy STRESS IN RESTRAINED COMPONENT II
FOR TEMPERATURE PROFILE 3
(STEADY-STATE)**

Figure 16.34 summarizes the maximum tensile stress locations and crack directions. The component II maximum tensile stresses are in general less than those of component I. However, component II develops two maximum tensile stress conditions along the radial joint. The second maximum tensile stress on the back side of the radial joint is identical with an asterisk. The unrestrained component II develops only one maximum tensile stress at the radial joint. The elastic-plastic behavior used for the restrained component II tends to cause higher maximum tensile stresses than those evaluated using only an elastic behavior.

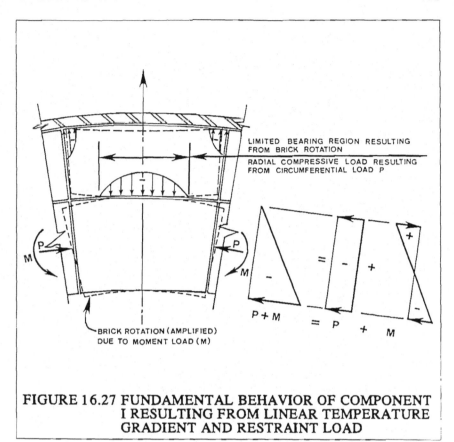

FIGURE 16.27 FUNDAMENTAL BEHAVIOR OF COMPONENT I RESULTING FROM LINEAR TEMPERATURE GRADIENT AND RESTRAINT LOAD

V. CONCLUSION

Based on the analytical results, the following statements are made regarding the influence of lining restraint and elastic-plastic behavior of the refractory lining material in predicting thermal shock fracture during rapid heating and cooling.

1. Heating the restrained refractory lining above the plastic strain threshold temperature results in compressive plastic straining in the hotface region causing amplification of tensile thermal stress during cooldown.

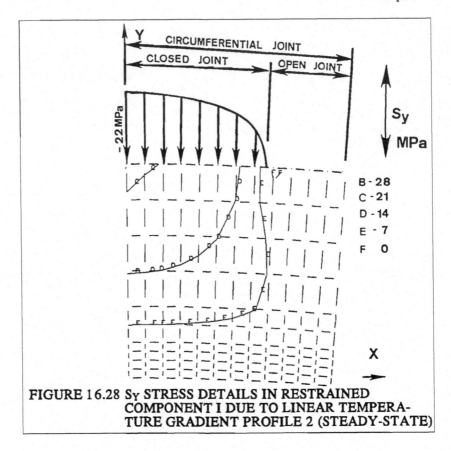

FIGURE 16.28 Sy STRESS DETAILS IN RESTRAINED
COMPONENT I DUE TO LINEAR TEMPERA-
TURE GRADIENT PROFILE 2 (STEADY-STATE)

2.   Generally, in the analytical evaluation of a restrained
        lining component, an elastic analysis solution will predict
        realistic critical heating rates. Realistic critical cooling
        rates can be predicted if the lining is not heated above
        the plastic strain threshold temperature.

3.   For heated restrained lining components of varying
        dimensions, lining restraint reduces thermal tensile
        stress fracture along the radial joint to a single
        circumferential direction. The single circumferential
        fracture is caused by the strain discontinuity at the tail
        of the circumferential restraint load along the radial
        joint.

RESTRAINT CONDITION	ANALYSIS TYPE	HEATUP TRANSIENT		STEADY-STATE HEATED	
		S_x, MPa	S_y	S_x	S_y
RESTRAINED		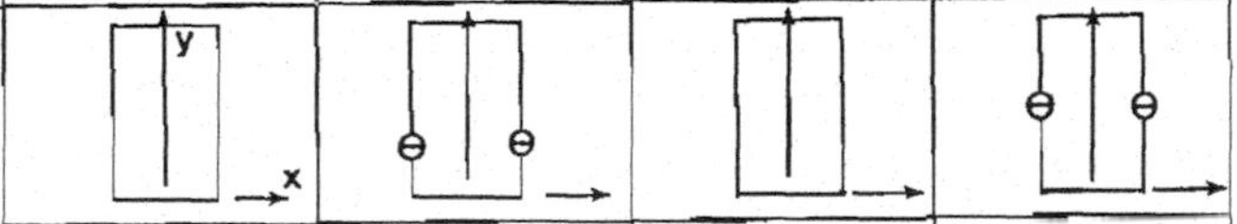			
	ELASTIC	0	17	0	28
	ELASTIC-PLASTIC	0	17	0	26
UNRESTRAINED		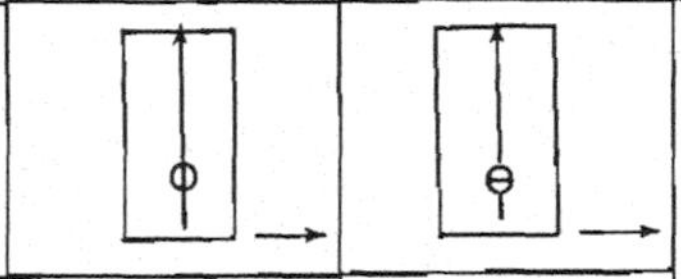			
	ELASTIC	18	20		

FIGURE 16.29 SUMMARY OF MAXIMUM THERMAL TENSILE STRESS IN COMPONENT II FOR HEATUP AND STEADY-STATE, PROFILES 1, 2 AND 3

4. During rapid cooldown of a restrained lining, maximum radial tensile stress (S_{ym}) will develop at the tail of the restraint load along the radial joint. The maximum circumferential tensile stress (S_{xm}) will occur at the mid-width location of the hotface. With an increase in the length of the refractory lining component, a tail is developed at each end of the restraint load, resulting in two local maximum radial tensile stresses along the radial joint.

5. In a restrained lining component, the amount of compressive plastic strain that occurs in the heated condition is a function of the magnitude of the lining component temperature that exceeds the plastic strain threshold temperature.

6. The maximum radial tensile stress (S_{ym}) along the radial joint is due to the strain discontinuity between the restrained and unrestrained surfaces of the radial joint. That is, if ϵ_U represents the unrestrained strain near the discontinuity and ϵ_R represents the restrained strain near the discontinuity:

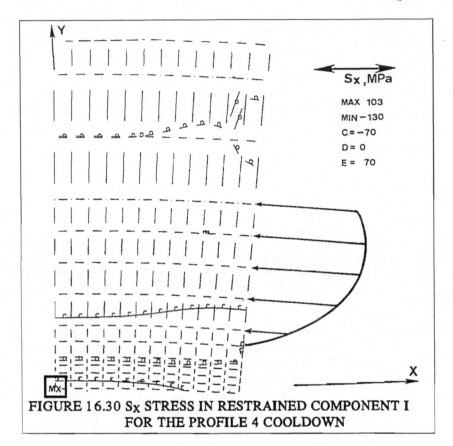

FIGURE 16.30 S_X STRESS IN RESTRAINED COMPONENT I
FOR THE PROFILE 4 COOLDOWN

$$\frac{d\epsilon_U}{dy} \neq \frac{d\epsilon_R}{dy}$$

Therefore, at the discontinuity:

$$\frac{d\epsilon_U}{dy} - \frac{d\epsilon_R}{dy} = f\left[\frac{d^2\epsilon}{dy^2}\right] > 0$$

A radial tensile thermal stress exists (where ϵ is the strain function at the discontinuity).

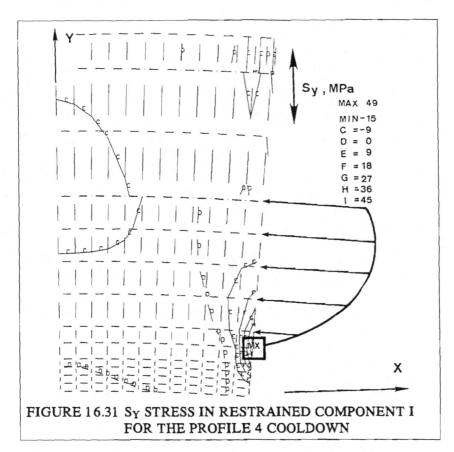

FIGURE 16.31 S$_Y$ STRESS IN RESTRAINED COMPONENT I
FOR THE PROFILE 4 COOLDOWN

7. The second maximum radial tensile stress (S$_{ym}$) at the interior tail of the restraint load along the radial joint, remote from the hotface, appears to be independent of the transient heating or cooling applied to the hotface. Also, this stress appears to be independent of the compressive plastic straining that occurs in the hotface region of the lining component. As a result, thermal spalling can occur due to the steady-state heating condition.

8. Maximum radial tensile stress (S$_{ym}$) developed at the tail of the restraint load along the radial joint near the hotface appears to be dependent on the amount of

RESTRAINT CONDITION	ANALYSIS TYPE	INITIAL COOLDOWN		SECOND COOLDOWN	
		S_x, MPa	S_y	S_x	S_y
RESTRAINED		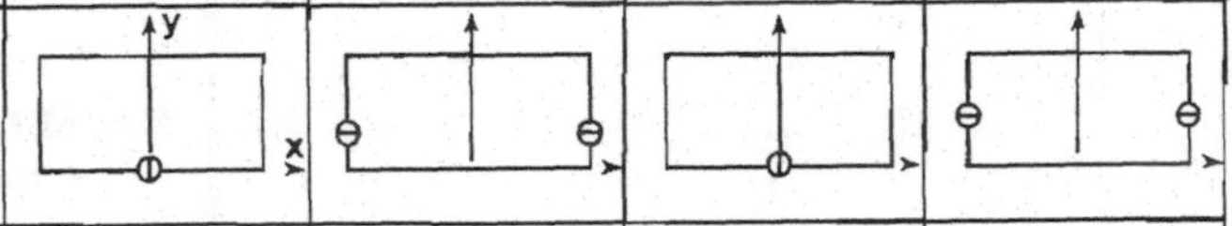			
	ELASTIC	33	3	116	63
	ELASTIC-PLASTIC	103	49	166	93
UNRESTRAINED		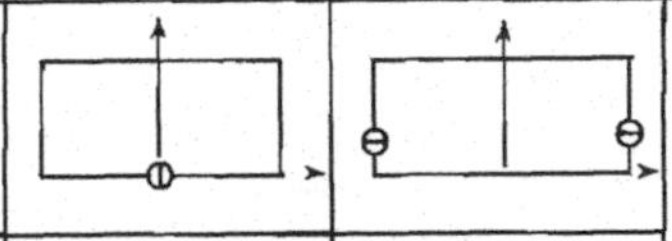			
	ELASTIC	94	47		

FIGURE 16.32 SUMMARY OF MAXIMUM THERMAL TENSILE STRESS IN COMPONENT I FOR COOLDOWN PROFILES 4 AND 5

compressive plastic straining and the cooling or heating rate.

9.   The maximum circumferential tensile stress at the mid-width of the lining component hotface appears to be dependent on the amount of compressive plastic straining and the cooling rate.

10.   The maximum tensile stress (S_m, maximum of S_x or S_y) developed in a restrained refractory lining component is dependent on many variables, expressed as:

$$S_m = f(t, \phi, a_B, \alpha_B, E_{EB}, E_{PB}, \upsilon_B, L_B, W_B, K_S, T_B, T_{PT}, \delta_B, dT_D/dy)$$

indicating that S_m is a function of time (t), heating or cooling rate (ϕ), refractory component thermal diffusivity (a_B), refractory component thermal expansion (α_B), refractory component elastic modulus (E_{EB}), refractory component plastic modulus (E_{PB}), refractory component Poisson's ratio (v_B), the refractory component length (L_B) and width (W_B), the stiffness of the vessel shell (K_S), the steady-state temperature of the brick component prior to heating or

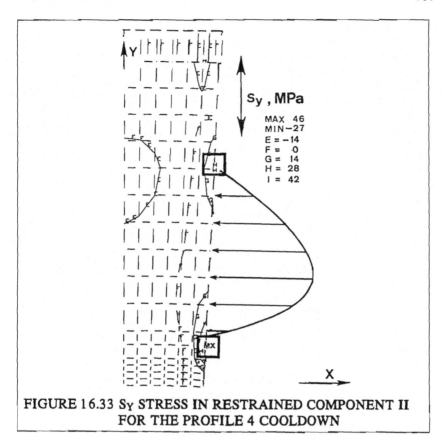

FIGURE 16.33 S_Y STRESS IN RESTRAINED COMPONENT II
FOR THE PROFILE 4 COOLDOWN

cooling (T_B), the plastic strain threshold temperature (T_{PT}), the expansion allowance used in the circumferential and radial joints surrounding the brick component (δ_B) and the slope of the temperature gradient at the point of the maximum radial tensile stress along the radial joint at the tail of the restraint load (dT_D/dy).

VI. SUMMARY

A. Material Tensile Strength

Refractory engineers use the chemistry of refractories,

RESTRAINT CONDITION	ANALYSIS TYPE	INITIAL COOLDOWN S_x, MPa	INITIAL COOLDOWN S_y	SECOND COOLDOWN S_x	SECOND COOLDOWN S_y
RESTRAINED	ELASTIC	30	26.7	77	17.54
RESTRAINED	ELASTIC–PLASTIC	76	30.46	97	15.72
UNRESTRAINED	ELASTIC	66	50		

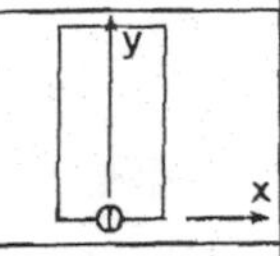

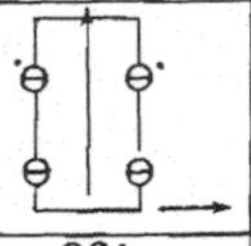

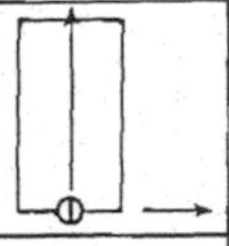

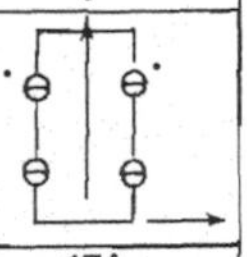

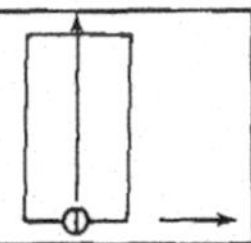

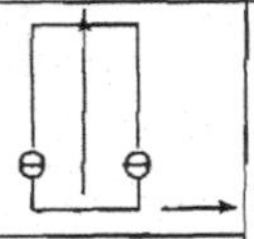

FIGURE 16.34 SUMMARY OF MAXIMUM THERMAL TENSILE STRESS IN COMPONENT II FOR COOLDOWN PROFILES 4 AND 5

processing procedures and other related manufacturing processes to maximize the strength of refractories and, more specific to our discussion, the tensile strength. The refractory component in the refractory lining system is exposed to two types of loads: stress-controlled loads and strain-controlled loads. Work of fracture (WOF) is a testing procedure for refractory materials measuring their strength. By maximizing the area under the WOF force-displacement curve, the strength is also said to be optimized. However, the load type to which the refractory is exposed is not considered as a part of the material strength optimization. Therefore, it is recommended that the parameters for material strength optimization be revised to:

I. For refractory exposed primarily to stress-controlled loadings:

    a. Maximize the area under the WOF force-displacement curve.

    b. Maximize the ultimate tensile stress, σ_f.

II. For refractories exposed primarily to strain-controlled loadings:

a. Maximize the area under the WOF force-displacement curve.

b. Maximize the ultimate tensile strain, ϵ_f.

B. Lining Tensile Stress

As shown in this investigative analysis, the cause of tensile thermal stress fracture and the resulting deterioration of the restrained lining component is highly complex and dependent on many variables. Fracture of the restrained lining component is not only dependent on transient heating and cooling rates, but also on the steady-state temperature gradient and the maximum temperature imposed on the component.

In general, the restraint applied to a lining component reduces maximum tensile stress during heatup. If the lining temperature does not exceed the plastic strain threshold temperature, restraint tends to reduce the maximum tensile stress during cooldown. If the lining temperature exceeds the plastic strain threshold temperature, the restraint tends to increase the maximum tensile stress in the region of the lining hotface during cooldown.

Pinch spalling deterioration at the hotface corners of the radial joint is caused by the shear stress produced by the restraint loading on the radial joint.

Slabbing deterioration appears to be caused by crack initiation due to maximum tensile stress developed at the radial joint surface. The maximum tensile stress causing slabbing might be developed during heatup, cooldown or during steady-state temperatures. In general, cooldown appears to be more critical than heatup in causing lining mechanical deterioration. The brick component with a narrow hotface width and a significant length appears to develop higher tensile thermal stress at the radial joints during steady-state temperatures than during transient heatup.

The results and conclusions of this study will, most likely, vary due to the rate of heating or cooling, the elastic-plastic behavior and other mechanical material properties of the refractory lining and the thermal material properties of the refractory lining.

Thermal tensile fracture is an extremely complex behavior and must be evaluated on a case-by-case basis.

REFERENCES

1. Kingery, W. D., Factors Affecting Thermal Stress Resistance of Ceramic Materials, Journal of American Ceramic Society, Vol. 38, No. 1, pp. 3-15 (1955).

2. Manson, S. S. and Smith, R. W., Theory of Thermal Shock Resistance of Brittle Materials Based on Weibull's Statistical Theory of Strength, Journal of American Ceramic Society, Vol. 38, No. 1, pp. 18-27 (1955).

3. Hasselman, D. P. H., Unified Theory of Thermal Shock Fracture Initiation and Crack Propogation in Brittle Ceramics, Journal of American Ceramic Society, Vol. 52, No. 11, pp. 600-604 (1969).

4. Kienow, V. S., Crack Formation in Fired Converter Bricks, Berichte der Deutschen Keramischen Gesellschaft, Vol. 48, No. 7, pp. 462-430 (1970).

5. Hasselman, D. P. H., Thermal Stress Resistance Parameters for Brittle Refractory Ceramics: A Compendium, American Ceramic Society Bulletin, Vol. 49, No. 12, pp. 1033-1037 (1970).

6. Larson, D. R. and Hasselman, D. P. H., Comparative Spalling Behavior of High-Alumina Refractories Subjected to Sudden Heating and Cooling, Transactions of British Ceramic Society, Vol. 74, No. 2, pp. 59-65 (1975).

7. Brezny, B. and Shultz, R., Determining the Thickness of Gunned Layer Needed to Protect BOF Brick from Thermal Shock Damage, Industrial Heating, April, 1978, pp. 16-17.

8. Shultz, R., Brezny, B. and Hambrick, D., The Effect of

Gunning on the Thermal Shock Resistance of Steel Ladle Refractories, Third International Iron & Steel Congress, Chicago, IL, April 16-20, 1978.

9. Ainsworth, J. H., Calculation of Safe Heat-Up Rates for Steel Plant Furnace Linings, American Ceramic Society Bulletin, Vol. 58, No. 7, pp. 676-678 (1979).

10. Brezny, B., Crack Formation in BOF Refractories During Gunning, American Ceramic Society Bulletin, Vol. 58, No. 7, pp. 679-682 (1979).

11. Freiman, S. W., Brittle Fracture Behavior of Ceramics, American Ceramic Society Bulletin, Vol. 67, No. 2, pp. 392-402 (1988).

12. Homeny, J. and Bradt, R. C., Thermal Shock Fracture, Thermal Stress in Materials and Structures in Severe Thermal Environments, (Hasselman, D. P. H. and Heller, R. A., eds.), Plenum Publishing Co., New York, pp. 343-364, 1980.

13. Bradt, R. C., Fracture Measurements of Refractories: Past, Present and Future, American Ceramic Society Bulletin, Vol. 67, No. 7, pp. 1176-1178 (1988).

14. Nakayama, J., Bending Method for Direct Measurement of Fracture Surface Energy of Brittle Materials, Jpn. J. Appl. Phys., 3 [7]: pp. 422-423 (1964).

15. Nakayama, J. and Ishizuka, M., Experimental Evidence for Thermal Shock Damage Resistance, American Ceramic Society Bulletin, Vol. 45, No. 7, pp. 666-669 (1966).

16. Bradt, R. C., Elastic Moduli, Strength and Fracture Characteristics of Refractories, Key Engineering Materials, Trans. Tech. Publication, Switzerland, Vol. 88, pp. 165-192, (1993).

17. Bradt, R. C., Thermal Fracture of Bosh Area Refractories, Progress Report AISI Proj. 46-339, AISI August 1, 1990.

18. Braiden, P. M., The Development of Rational Design Criteria for Brittle Materials, Materials In Engineering, Vol. 2, pp. 73-82, Dec., 1980.

19. Hasselman, D. P. H., Thermal Stress Resistance Parameters for Brittle Refractory Ceramics: A Compendium, American Ceramic Society Bulletin, Vol. 49, No. 12, pp. 1033-1036 (1970).

20. Hasselman, D. P. H., Elastic Energy at Fracture and Surface Energy as Design Criteria for Thermal Shock, Journal of American Ceramics Society, Vol. 46, No. 11, pp. 535-540 (1963).

21. Hasselman, D. P. H., Unified Theory of Thermal Shock Fracture Initiation and Crack Propogation in Brittle Ceramics, Journal of American Ceramics Society, Vol. 52, No. 11, pp. 600-604 (1969).

22. Bradley, F., Chaklader, A. C. D., and Mitchell, A., Safe Heating and Cooling Rates as Theoretical Criteria for the Design and Selection of Refractory Structural Components, 1985 AISE Annual Convention, Pittsburgh, PA, Iron and Steel Engineer Abstracts, Vol. 62, No. 7, p. 74, July, 1985.

23. Bradley, F., Chaklader, A. C. D. and Mitchell, A., Thermal Stress Fracture of Refractory Lining Components: Part I Thermoelastic Analysis; Part II Safe Heating and Cooling Rates; Part III Analysis of Fracture, Metallurgical Transactions B., Vol. 18B, No. 2, pp. 355-380, June, 1987.

24. Bradley, F., Chaklader, A. C. D. and Mitchell, A., Theoretical Criteria for the Design and Selection of Refractory Structural Components, Department of Metallurgical Engineering, University of British Columbia, Vancouver, British Columbia, Canada, circa 1988.

25. Fujiwara, A. and Fujino, M., Stress Analysis of Axisymmetrical Refractory Structures Under Restraint During Thermal Expansion, Tetsu-to-Hagane, Vol. 70, No. 2, pp.

208-215 (1984).

26. Schacht, C. A., Influence of Lining Restraint in Predicting Thermal Shock Fracture of Refractory Linings, American Society of Ceramic Engineers 91st Annual Meeting, Indianapolis, IN, April 23-27, 1989.

27. Schacht, C. A., Influence of Lining Restraint and Non-Linear Material Properties in Predicting Thermal Shock Fracture of Refractory Linings, American Ceramic Society, Unitcer '89 Proceedings, Vol. 2, pp. 1286-1316, 1989.

28. Kato, I., Morita, Y. and Hikami, F., Studies on Thermal Spalling of Refractories, Central Research Laboratories, Sumitomo Metal Industries, Ltd., Research Report, pp. 32-661 to 32-667.

29. Kato, I., Morita, Y. and Hikami, F., Thermal Stress Formula for Estimation of Spalling Strength of Rectangular Refractories, Tetsu-to-Hagane, Vol. 68, No. 1, pp. 105-112 (1982).

30. Kumagai, M., Vchimura, R. and Emi, T., Factors Effecting Thermal Shock Damage of Fireclay Brick, Refractory Material Laboratory, Research Laboratories, Kawasaki Steel Corp. Research Report, pp. 32-601 to 32-608.

31. Uchiyama, S., Views on Fracture Behavior of Refractories, Source Unknown (from Japanese Steel Industry), pp. 32-539 to 32-547.

32. Jacquemier, M., Study of Developed Constraints in Refractory Masonry, French Ceramic Society, pp. 1-37, June 5, 1981.

33. Coatney, R. L., Stress Analysis of a Brick Lined Rotary Kiln, American Ceramic Society, presented at Refractory Division Meeting, Bedford Springs, PA, Oct. 5, 1979.

34. ANSYS User's Manual, Swanson Engineering, Systems, Inc., Houston, PA, Rev. 4.4, May 1, 1989.

35.   Carslaw, H. S. and Jaeger, J. C., Conduction of Heat in Solids, Oxford University Press, London, England, 1959, p.63.

17
Results of Investigative Studies on Various Industrial Refractory Systems

I. INTRODUCTION

The objective of this chapter is to provide details of various investigative studies [1] on refractory-lined industrial vessels and other industrial refractory structures. In some cases, strain gauges and thermocouples were used to verify analytical predictions. In some cases, vessel inspections were made to verify the reported refractory lining problems. In other cases, discussions were conducted with engineers who observed the refractory lining behavior and were able to provide details of their observations.

The reason for the investigative studies varied from simply confirming expansion behavior of the lining system to identifying the cause of refractory deterioration.

II. REFRACTORY SPRUNG ARCH

A. Description of Sprung Arch

The refractory sprung arch has been a popular cover for tunnel kilns and other industrial refractory structures. The basic components [2] of the sprung arch are described in Figure 17.1. The external structural support consists of continuous horizontal beams along the exterior end of each skew. Vertical columns (called buckstays) are spaced at uniform intervals along the length of the arch. The spacing of these vertical columns are a function of the horizontal beam stiffness. The horizontal thrust of the arch will cause the horizontal beams to displace. The spacing of the vertical columns is selected to maintain a reasonable displacement in the horizontal beam.

The base of each vertical column is designed as a pin connection, allowing the column to rotate from the base without resistance. The top ends of each pair of vertical columns are connected by a rod and spring system. The spring stiffness is selected to be compatible with the expected horizontal thrust, F_H (see Chapter 12). The spring can be positioned by the thread/nut system at the end of the rod which supports the spring. As the arch is heated, the nut is backed off to allow for the appropriate horizontal outward movement of the horizontal beam and skew brick and the resulting expansion allowance for the arch.

The primary load-carrying arch is shown as arch brick components. The insulating brick, block and blanket materials are shown as a cross-hatched pattern over the top of the arch.

The arch investigated had the following dimensions (see Figure 17.2) [3,4]:

radius, r	=	2800 mm
angle, $\phi/2$	=	23° 31'
rise, h	=	232 mm
thickness, t	=	228 mm
span, s	=	2235 mm

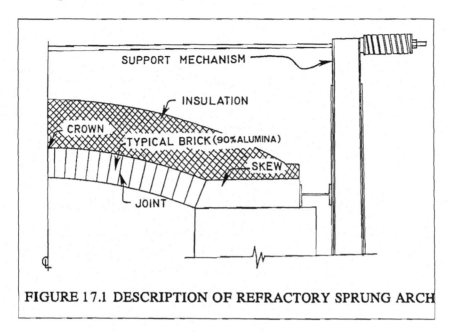

FIGURE 17.1 DESCRIPTION OF REFRACTORY SPRUNG ARCH

The insulating materials, over the top of the arch, were defined as 342 mm of insulating brick, consisting of three layers, 114 mm thick, and a 25 mm thick insulating blanket material placed over the top outside surface of the insulating brick. The arch was constructed of fired 90% alumina arch brick.

As a matter of interest, the arch thickness is checked against the Equation (12.7) criteria. Basically, the arch thickness should be less than:

$$t < \{h - [1/48S]\}/\cos 0/2 \qquad (17.1)$$

or:

$$t < \{232 - [1/(48)(2235)]\}/\cos 23.52 \ = \ 253 \text{ mm}$$

Since t = 228 mm < 253 mm, the arch geometry is satisfactory.

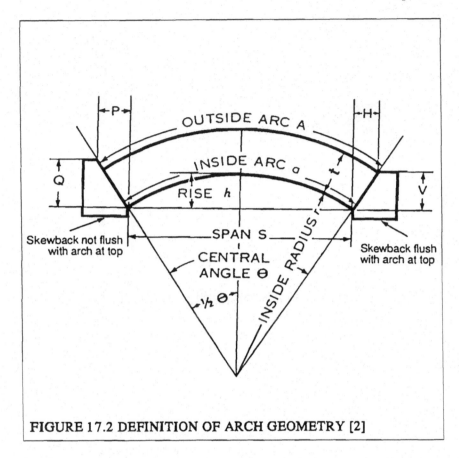

FIGURE 17.2 DEFINITION OF ARCH GEOMETRY [2]

This criteria is not applicable for arches with full compensation for expansion. For arches with partial or no expansion compensation, the preceding relationship is applicable.

B. Arch Operating Environment

The arch hotface was exposed to a process temperature of 1700°C. The heat transfer analysis revealed that the arch coldface temperature was 1600°C. Because of the amount of insulation, the through-thickness temperature gradient was quite low. The arch was constructed as part of a tunnel kiln. However, the arch experienced considerable difficulties after only a few years of

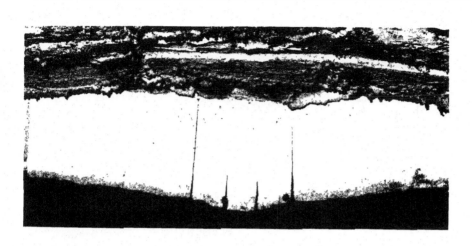

FIGURE 17.3 LOCALIZED INELASTIC REFRACTORY FLOW
AT ARCH CROWN HOTFACE

operation. The arch did not completely collapse, but slumped
inward into the tunnel kiln, as shown in Figure 17.3. Note the
excessive deformations at the hotface end of the radial joints in the
crown region of the arch. This reflects localized high compressive
stresses in the hotface end of the crown radial joints. The section
of the slumped arch is shown behind a portion of the arch that
remained stable. The irregular surface at the top of the photo is the
hotface portion of the closer part of the arch that remained stable.
The result was that plastic deformations occurred at these regions of
the joints. An investigative analysis was conducted in an attempt to
replicate the thermomechanical behavior of the arch and to provide
detailed insight into the arch behavior.

C. Sprung Arch Analysis

The arch was analyzed using the finite element method. A
finite element program was developed with constant strain triangle
as the basis of the single element. In addition, non-linear interface
elements were used at the radial joint to simulate a "compression
only" stress state for the radial joints. The model is illustrated in

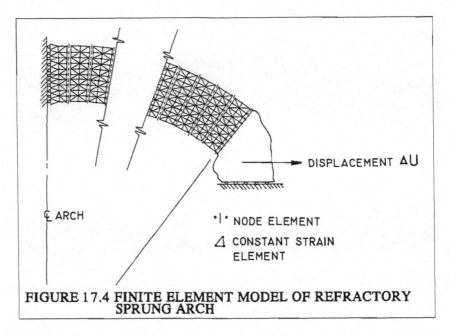

FIGURE 17.4 FINITE ELEMENT MODEL OF REFRACTORY
SPRUNG ARCH

Figure 17.4. Because of symmetry, only half of the arch was required for analysis.

The temperature-dependent modulus of elasticity of the arch brick was defined as:

$$MOE = 1.50 \times 10^2 - 0.0706T, \text{ GPa}$$

Because of the limited MOE data on alumina refractory materials at the time of this study, the above MOE equation was the best fit for sonic MOE test data, as described in Figure 17.5. The coefficient of thermal expansion was:

$$\alpha = 7.92 \times 10^{-6}, \text{ m/m°C}$$

Poisson's ratio for the material was:
$$v = 0.20$$

The average temperature of the arch was:

$$T_{AVE} = (1700 + 1600)/2 = 1650°C$$

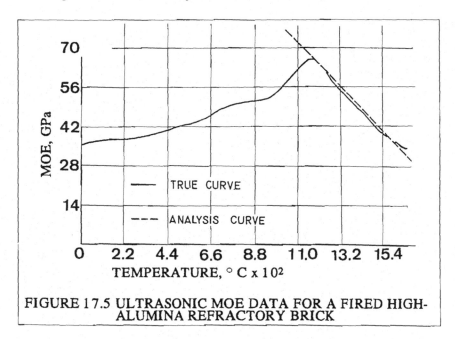

FIGURE 17.5 ULTRASONIC MOE DATA FOR A FIRED HIGH-ALUMINA REFRACTORY BRICK

Because only a half-span section was required for the analysis, the expansion allowance (Δu) was evaluated as:

$$\Delta u = \alpha S \Delta T / 2 \qquad (17.2)$$
$$= (7.92 \times 10^{-6})(2325)(1650)/2$$
$$= 15 \text{ mm}$$

where S is the span at the arch centerline. The insulating material gravity load was applied as a uniform load of 90 N/m along the full span of the arch.

The arch elastic analysis results are best described by the radial joint behavior as shown in Figure 17.6. Hinges are formed at the center and two skew regions. However, the joints actually undergo a transition from partial bearing of the joints at the hinges to full bearing of the joints in regions adjacent to the hinges.

As expected, the highest stresses exist at the three hinge points. Figure 17.7 describes the circumferential compressive stresses in the arch. The high stresses at the hinge points will cause

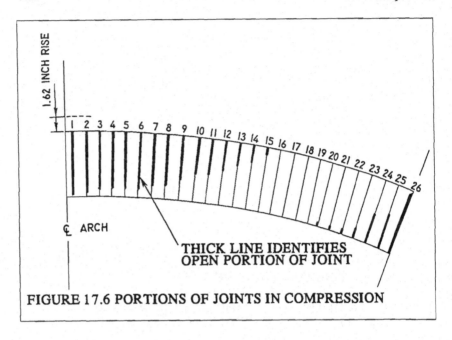

FIGURE 17.6 PORTIONS OF JOINTS IN COMPRESSION

plastic deformations at these points. The analytical results are consistent with the observed inelastic deformations described in Figure 17.3.

D. Conclusions

The finite element solution of the refractory brick arch is path-dependent. The behavior of the model is complex and highly sensitive to the rate at which the loads are applied in the analysis, and care must be exercised in evaluating the refractory structures. However, the elastic solution using the finite element technique agrees well with experimental data on a multijointed arch constructed of steel blocks. The analysis of a refractory sprung arch revealed an interesting behavior and has practical application.

From the analysis of the sprung arch it appears that several critical regions merit attention. At initial conditions, before significant creep flow has begun, the brick at the skew and near the crown of the arch experience the highest stresses. Even with movement of the skew for thermal growth the arch still exhibits

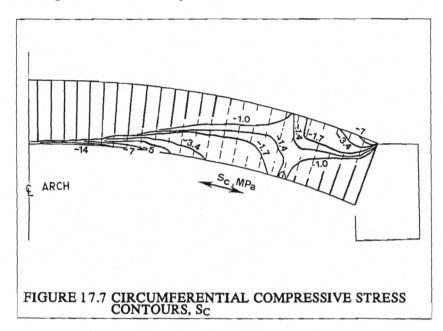

FIGURE 17.7 CIRCUMFERENTIAL COMPRESSIVE STRESS CONTOURS, Sc

high stresses because of the small bearing surface at the joints. It seems reasonable that the more dense, higher-strength brick be placed in these regions of high stress.

Many other ideas have been used to absorb the thermal expansion forces in other types of refractory arches. These ideas have included inserting cardboard or asbestos liners in the joints or the use of special designs of metal-encased brick. This analysis indicates that determining the location of joint openings may be complex. For the joint opening condition considered in this analysis, the expansion strips would be ineffective on the hotface near the skew and the coldface of the crown since no contact occurs at these locations immediately after heatup.

The joint opening distribution is related to the amount of expansion allowance. In this respect, careful consideration should be given to the design of a thermal expansion mechanism if it is to be used successfully.

Caution should be used when applying this solution to other

types of refractory arches or to those cases where different thermal expansions, materials and boundary conditions are used in their design.

III. SPHERICAL REFRACTORY SILICA BRICK DOME

A. Description of Spherical Silica Brick Refractory Dome

An analysis was conducted on the spherical refractory dome shown in Figure 17.8. The refractory spherical dome described is part of a complex refractory lining system used in a blast furnace stove. The blast furnace is part of the equipment used in the pro-

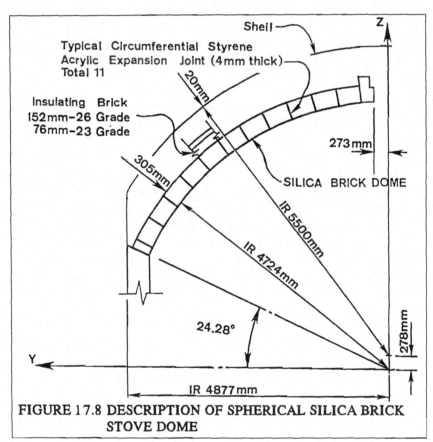

FIGURE 17.8 DESCRIPTION OF SPHERICAL SILICA BRICK STOVE DOME

duction of pig iron in a steel mill. The stove is used to conserve heat from spent gas as well as to heat the fresh supply gas used in the blast furnace operation, which reduces iron ore to pig iron [5]. This portion of the steel mill is referred to as the ironmaking, or primary, portion of the steel mill. The steelmaking portion of the steel mill converts the pig iron to steel by adding alloys and removing impurities and other undesirable chemical components.

The blast furnace stove dome described in Figure 17.8 was constructed with a silica brick work lining. Typically, the stove dome work lining is constructed of fired alumina brick. For high-temperature gas usage in the blast furnace, the silica brick is used because of the greater silica brick strength at higher temperatures. Higher gas temperatures are used to enhance the process of converting iron ore to pig iron in the blast furnace.

As described in Figure 17.8, the dome had an inside radius of 4724 mm and a silica brick work lining thickness of 305 mm. The skew face angle was 24.78° measured up from the horizontal. Insulation was used on the cold side of the dome that had a total thickness of 228 mm.

The steel shell is 20 mm thick in both the cylindrical and spherical portions. The dome had a radius of 5500 mm and the cylindrical part a 4877-mm radius.

The silica brick expansion curve is shown in Figure 17.9 [6]. As expected with silica brick, the rate of expansion is highly non-linear and is temperature dependent. The greatest amount of expansion takes place in the temperature range between room temperature and about 300°C.

The primary concern of the silica brick stove dome design with regard to the amount of expansion allowance is whether sufficient expansion allowance is used to prohibit overstressing the silica brick during heatup and resulting stress fracture. The compressive failure stress for the silica brick was established at 34 MPa.

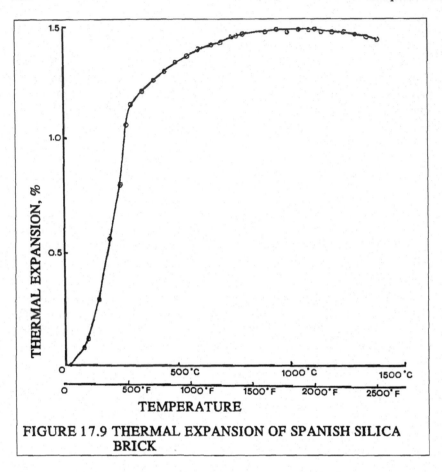

FIGURE 17.9 THERMAL EXPANSION OF SPANISH SILICA
        BRICK

B. Background on a Silica Brick Stove Dome
Experiencing Spall Fracture

The concern over stress fracture of the silica brick stove
dome is   valid. Another silica brick blast furnace stove dome of
similar size experienced stress fracture during heatup. The fracturing
occurred on the inside face of the dome at about the seventh course
up from the skew, as illustrated in Figure 17.10. The spall fracture
was noticed when the hotface combustion gas reached about 540°C.
The calculated angular location of the spall was at an angle of
about 14° up from the skew face.

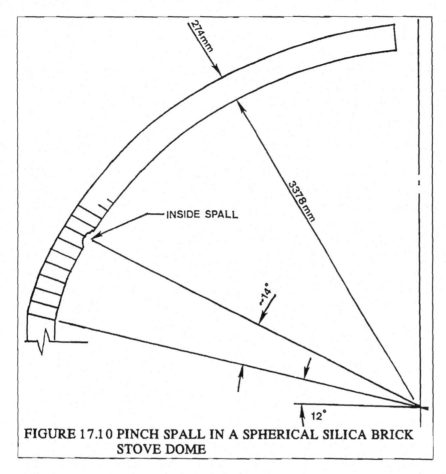

FIGURE 17.10 PINCH SPALL IN A SPHERICAL SILICA BRICK
STOVE DOME

The strength data on the silica brick used in this dome showed a MOR strength of 5.5 MPa and a CCS of 38.50 MPa. The silica brick thermal expansion was found to be 1.5% at 1100°C. The dome was designed with an expansion allowance for a silica brick with an expansion of 1.3% at 1100°C.

Cardboard was used as an expansion allowance material. The cardboard was defined as having a 2.5-mm thickness. A single layer of cardboard was used at every second circumferential joint throughout the full dome. The alternate joints were mortar joints.

In the first six courses (measured up from the skew), every

five vertical joints consisted of a single layer of cardboard in four of the five joints. A mortar joint was used in the fifth joint. In the remaining courses of brick (from seventh course up to top of dome), the vertical joint expansion allowance consisted of a single layer of cardboard in every third vertical joint. The remaining vertical joints in these courses consisted of mortar joints.

The percent expansion allowance was based on the assumption that 80 percent of the cardboard burned out at 200 to 260°C. Therefore, the cardboard thickness used for expansion allowance was 2 mm (0.80 x 2.4 = 1.92). Therefore, at temperatures below 200 to 260°C, no expansion allowance was available. However, the expansion forces are only a small fraction of the values that would exist at full expansion. Based on a 1.5% brick expansion at 100°C, Table I summarizes the percent expansion allowance in the various dome joints. As shown in Table I, an abrupt change in expansion allowance occurred in the vertical joints of the sixth and seventh courses. That is, the circumferential expansion stresses in the silica brick would experience an abrupt change at the sixth and seventh rows.

The results of the analysis of the initial silica brick stove dome will be used to explain why the spall fracture occurred at the location shown in Figure 17.10. As previously discussed in Chapter 14, a hinge occurs at the location of the spall. Although an analysis was not conducted on this silica brick stove dome, the reason for the spall fracture is a result of thermomechanical events unfortunately occurring at the same location in the dome. That is, the abrupt change in circumferential expansion stresses occurred at the same location as the upper hinge. As a result, high stresses and high-stress gradients existed at the skew, resulting in a circumferential spall fracture, as illustrated in Figure 17.10.

C. Theoretical Aspects in Brick Stove Dome Design

The theoretical aspects in brick dome designs are presented for the catenary and spherical stove domes. For the gravity load condition, this dome shape has considerably lower stresses at the skew region than the spherical shaped dome. However, the gravity

Table I

Description of Expansion Allowance in
Spall Fractured Silica Brick Stove Dome

Joints	Course No. Measured Up From Skew	% Expansion Allowance
Circumferential Joints	All of Dome	58
Vertical Joints	1 to 6	100
	7 to Top	45

load stresses are usually much smaller than the thermal expansion stresses. Therefore, the choice of dome design must be made based on the stresses developed during the hot condition. The amount of expansion allowance determines the stresses during the hot condition and is a primary parameter for evaluating the dome design and not the dome geometry. The type of refractory brick used has a direct influence in the design considerations. Silica brick has higher expansion rates than fireclay and high-alumina brick. The expansion allowance is of primary concern in both the catenary and spherical dome constructed of silica brick.

When considering only the gravity load condition, the catenary type of dome does not develop significant bending stresses at the base of the dome near the skew. The spherical dome, however, develops higher bending near the skew for the gravity load condition. Since gravity load stresses are considerably less than the thermal expansion stresses, the gravity load stresses are of secondary concern. The primary concern is the amount of expansion allowance used to avoid the development of excessive thermal expansion stresses and failure of the refractory.

The most popular method of incorporating expansion allowance into the dome design is to place thin sheets of combustible material in the brick joints. This material burns out during heatup. Typically, the combustible material is placed in both the vertical and horizontal brick joints. The combustible material in the vertical joints provides expansion allowance in the circumferential direction, as shown in Figure 17.11. The combustible material in the circumferential joints (between various courses of brick) provides expansion allowance for both the vertical and horizontal directions of expansion. The horizontal component (T_H) of the circumferential joint expansion allowance adds to the vertical joint expansion allowance. That is:

$$\Delta'_C = 2\pi T_H + \Delta_C \tag{17.3}$$

where

Δ'_C = Total effective circumferential expansion allowance at a given horizontal brick course in the dome

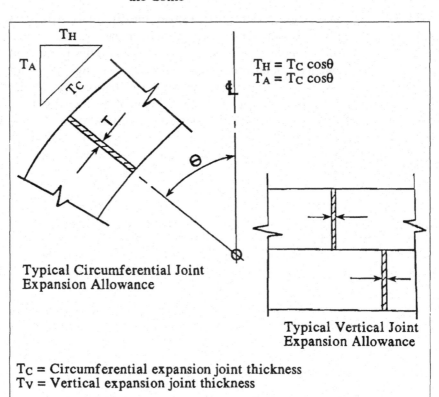

Typical Circumferential Joint
Expansion Allowance

$T_H = T_C \cos\theta$
$T_A = T_C \cos\theta$

Typical Vertical Joint
Expansion Allowance

T_C = Circumferential expansion joint thickness
T_V = Vertical expansion joint thickness

FIGURE 17.11 HORIZONTAL AND VERTICAL JOINT
EXPANSION ALLOWANCE

Δ_C = Total circumferential expansion allowance in vertical joints

Δ_C = ΣT_v

$2\pi T_H$ = Circumferential expansion allowance provided by the horizontal component (T_H) of the circumferential joints. This is for a circumferential joint located at angle θ measured from the dome axis.

$2\pi T_H$ = $2\pi T_C \cos\theta$

The total circumferential expansion allowance for the circumferential brick course at angle θ is:

$$\Delta'_C = 2\pi T_C \cos\theta + \Sigma T_v \qquad (17.4)$$

The expansion allowance material used in the vertical joints provides considerably more circumferential expansion allowance than the circumferential joints for any given course of brick. At any given brick course, each circumferential joint expansion allowance material provides the total circumferential expansion allowance, described as:

$$2\pi T_H = 2\pi T_C \cos\theta$$
$$= 6.28\ T_C \cos\theta$$

The effective thickness (T_C) of the circumferential joint burnout material is amplified by 2π, but reduced by $\cos\theta$.

The total number of vertical joints with burnout material, in the lower regions of the dome, may be about 60 to 70. Therefore, the total circumferential expansion allowance from only the vertical joints in the lower regions of the dome would be about $60T_C$. If the vertical joint and horizontal joint burnout material is assumed to be the same thickness ($T_v = T_C$), then the vertical joints have considerably more expansion allowance material than the horizontal joints by the ratio (using $\theta \approx 60°$):

$$= \frac{60T_v}{2\pi T_C \cos(60)} \qquad\qquad (17.5)$$

Since $T_v = T_c$:

$$= \frac{60}{2\pi(0.5)} \approx 20$$

Since the only restraint to the dome is the horizontal restraint imposed by the skew, the total circumferential component of the expansion allowance is of primary concern. In the lower region of the dome, near the skew region, the circumferential joints of the dome have less of a slope (θ is larger) than the circumferential joints in the upper part of the dome. Assuming that all these joints are keyed in some manner to prevent shear displacement (see Figure 17.12), the circumferential joints in the lower region of the dome provide less contribution to expansion allowance in the horizontal direction (see Figure 17.11) than the vertical joints. Thus, the circumferential joints in the lower region of the dome provide less effective circumferential expansion allowance than vertical joints. That is, the primary portion of the circumferential expansion allowance, in the lower region of the dome, is provided by the vertical joints. It should be noted that in the lower dome region a catenary dome has considerably less slope in the horizontal joints than the spherical dome. Therefore, more burnout material would be required in the lower dome vertical joints of a catenary dome than in the lower dome vertical joints of a spherical dome.

Since the catenary and spherical domes are not restrained in the vertical direction, the axial component (T_A) of the expansion allowance in the circumferential joints offers no assistance in reducing thermal expansion stresses. The axial component, however, is part of the expansion component of the circumferential joints and cannot be avoided. This implies that if a 100 percent circumferential expansion allowance is incorporated into the vertical joints, then no expansion allowance is required in the circumferential joints. The purpose of the additional amount of expansion allowance in the horizontal joints would be to preserve the geometry of the brick system, such that the hot geometry is

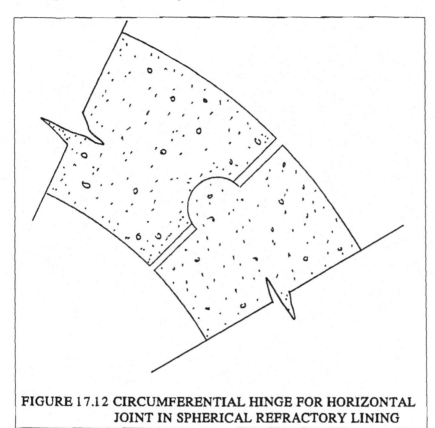

FIGURE 17.12 CIRCUMFERENTIAL HINGE FOR HORIZONTAL
            JOINT IN SPHERICAL REFRACTORY LINING

nearly identical to the cold geometry. By preserving the cold geometry, the development of hinges is minimized.

The expansion allowance for the circumferential expansion is more important for the lower portion of a dome than for the upper region of a dome. The skew, in most cases, provides the horizontal restraint to the dome at the base of the dome. The top crown portion of the dome is not restrained in the horizontal direction and, therefore, requires less expansion allowance. Shear lag, bending and the formation of hinges in the skew region dissipate and isolate the horizontal restraint of the skew from the upper regions of the dome.

In both catenary and spherical domes, the circumferential joints have a greater slope in the upper regions of the dome. These upper circumferential joints have a larger horizontal component of expansion allowance and, therefore, have a greater contribution to the circumferential expansion allowance in the upper regions of the dome. Typically, a constant amount of expansion allowance in the vertical joints is used throughout the full dome height. However, there is a greater need in the lower dome region for expansion allowance in the circumferential direction. That is, more of the circumferential expansion allowance is needed in the lower region of the dome due to the skew restraint.

The needs for expansion allowance in brick stove domes and other portions of the stove brick system should be carefully evaluated, and the theoretical considerations should be carefully understood. Silica brick systems require special considerations. It should also be noted that the catenary and spherical dome designs cannot be compared based on similar amounts of expansion allowance.

D. Material Properties and Operating Conditions of Spherical Silica Brick Stove Dome

The silica brick spherical stove dome was introduced in the Section III.A of this chapter.

The expansion allowance material was styrene acrylic pads 4 mm thick. This plastic material is supposed to completely burnout at about 95 to 120°C. Unlike the previous dome expansion allowance described in Section III.B, this dome had the same pattern of expansion allowance pads in the joints from the skew to the crown of the dome. Table II describes the expansion allowance using a 1.5% expanding silica brick at 1100°C.

The 4-mm thick plastic material was used in every third vertical joint throughout the full dome and every fourth circumferential joint throughout the full dome. The vertical joint expansion allowance resulted in a 4 mm expansion allowance every 435 mm (3 bricks x 145-mm width) and the circumferential expansion allowance resulted in a 4 mm expansion allowance every

440 mm (4 bricks x 110-mm length). Therefore, the expansion allowance for a 1.5% expansion of the silica brick at 1100°C is:

$$\text{Vertical Joints} = \frac{4 \text{ mm}}{435 \times 1.5\%} \times 100^2 = 61\%$$

$$\text{Circumferential Joints} = \frac{4 \text{ mm}}{440 \times 1.5\%} \times 100^2 = 61\%$$

The skew brick in the vertical radial joints used a 4-mm thick plastic pad every 7 bricks, or about every 965 mm. The skew brick had an expansion allowance for the circumferential direction of 28% $[(4/965 \times 1.5)(100)^2]$.

The stress-strain data [6] for the silica brick used to construct the work lining of the spherical dome are described in Figure 17.13. The stress-strain data reflect a change in stiffness over the temperature range of 300 to 1400°C. Starting from 300°C, the silica brick material stiffness increases when the temperature is increased to 600°C. At temperatures above 600°C (900, 1200 and 1400°C), the material continues to soften. A plot of the MOE calculated from the initial tangent of the stress-strain curves is tabulated in Table III.

Both the hot and cold condition have to be evaluated in a stove design. The evaluations should include the stresses and displacements of the dome during both the hot and cold conditions. Typically, the geometry of the dome and other portions of the stove are designed for the cold condition. During heatup the geometry is altered because of thermal expansion. The compatibility of the geometry with the hot condition is based on the amount of expansion allowance. With 100 percent expansion allowance, the geometry should be compatible with the hot condition, and stresses should be similar to the cold deadload condition. However, during subsequent cooldown the resultant geometry is not compatible and hinges will develop resulting in higher cooldown stresses. Lesser amounts of expansion allowance would result in higher stresses in the hot condition. With lesser amounts of expansion

Table II

Description of Expansion Allowance
in Successful Silica Brick Stove Dome Design

Joints	Courses	%Expansion Allowance
Circumferential Joints	All	61
Vertical Joints	All	61

allowance, the system would have less tendency to develop hinges, thereby avoiding the higher stresses at the hinges during cooldown. This also implies that with lesser expansion allowance, the geometry is more compatible with the cold condition resulting in less movement of the brick work during cooldown. According to discussions with plane engineers, they have seen stoves with excessive brick movement at cooldown and found it necessary to realign the brick work before the next heatup.

In summary, it appears that 100 percent expansion allowance may be desirable in minimizing expansion stresses in the hot condition. However, 100 percent expansion allowance would result in excessive brick work movement at cooldown. A prudent design would have sufficient expansion allowance to avoid excessive expansion stresses at hot conditions and avoid excessive brick work movement at cold conditions. If a stove system can be maintained at a hot condition without cooldown, then 100 percent allowance appears ideal since brick work movement in the cold condition would be avoided.

E. Results of Investigative Field Testing Analysis of
Spherical Silica Brick Refractory Dome

A description of the spherical brick refractory dome was previously provided in Section III.A. The following discussion addresses both the investigative field tests and the related investigative analysis on this refractory dome.

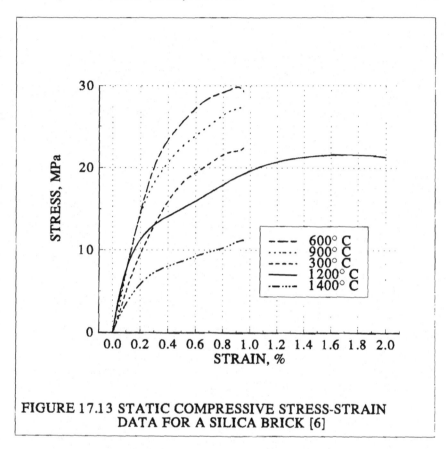

FIGURE 17.13 STATIC COMPRESSIVE STRESS-STRAIN
DATA FOR A SILICA BRICK [6]

Considerable data were developed from the 42 strain gauges and the 9 temperature gauges attached to the stove shell. Additional temperature gauges were attached to several locations within the brick work. A 28-day (about 672 hours) heatup was scheduled for the stove to take the stove temperature from room temperature up to about 1100°C. The primary interest here is the upper region of the shell where the skew and adjacent wall, the silica brick dome and the shell dome temperature all contribute to the development of stresses in the upper region of the shell at the location adjacent to the skew (see Figure 17.14).

The arrangement of the strain gauges and thermocouple at location A is shown in Figure 17.15. There were four sets of

Table III

Material Properties of Silica Brick
for Stove Dome Design

Temperature °C	Density* lb/cu ft	Poisson's Ratio	Modulus of Elasticity psi x 10^5	Coefficient of Expansion in./in.°F x 10^{-6}
70	142	0.2	0.5	7.04
600	142	0.2	1.0	23.00
1000	142	0.2	1.8	15.40
1400	142	0.2	2.1	11.07
2000	142	0.2	2.5	7.67
2600	142	0.2	2.2	5.77

*Effective density to include weight of insulating brick.

gauges positioned along a vertical line. Each set had a vertical gauge and horizontal (circumferential) gauge.

The strain gauge stress measurements at location A (top set gauges S_{m1} and S_{c2}) are plotted in Figure 17.15. These gauges were at an elevation on the vessel shell near the elevation of the skew. These results show that maximum shell stresses during heatup were achieved at about 300 hours. The maximum circumferential stress of nearly 40 MPa is reached. The maximum vertical stress is approaching -60 MPa. At 300 hours, the hotface of the dome brick had reached about 660°C. The coldface of the dome silica brick was at 585°C. Based on the dome temperature the stress-strain data (Figure 17.13) and the thermal expansion data (Figure 17.9), the maximum shell stress at 300 hours (Figure 17.16) seems very reasonable. At these dome temperatures, the silica brick has reached greatest stiffness in combination with the near greatest expansion. Because the heatup is very gradual, no significant transient stress effects are developed. This means that near steady-state temperatures exist at any point in time during heatup.

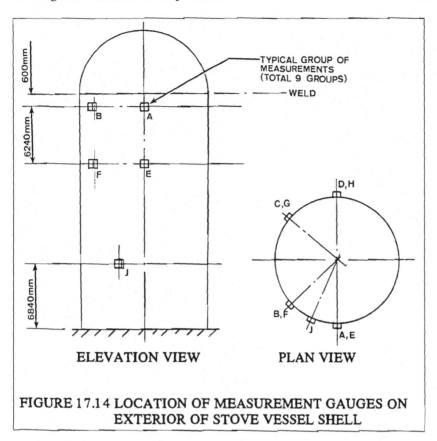

ELEVATION VIEW   PLAN VIEW

FIGURE 17.14 LOCATION OF MEASUREMENT GAUGES ON EXTERIOR OF STOVE VESSEL SHELL

Figures 17.17a and 17.17b are plots of the vertical (S_m) and circumferential (S_c) stresses at the two locations (A and B, respectively) around the regions of the skew in the upper shell wall near the dome shell. In each location, the circumferential stress (S_c) is continuously increasing in the vertical direction at the skew region. The vertical stress (S_m) represents the compressive vertical bending stress on the outside shell wall surface due to the outward loading of the skew. The gauges were at the outside surface of the shell. Therefore, since the S_m is compressive, the shell would be bending inward. This shell behavior, although it appears unrealistic, will be fully explained by the following analytical results.

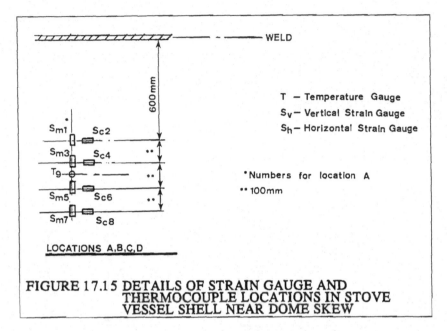

FIGURE 17.15 DETAILS OF STRAIN GAUGE AND
THERMOCOUPLE LOCATIONS IN STOVE
VESSEL SHELL NEAR DOME SKEW

The investigative analysis of the spherical silica brick refractory dome was performed using the finite element method [7]. An isometric view of the finite element model outline is described in Figure 17.18. A small pie section of the refractory dome was chosen for the model. Because of symmetry, a pie section is valid. However, the non-linear *compression only* stresses of the brick joints require appropriate joint elements to simulate this behavior in both of the primary directions within the dome. A pure axis symmetric model would not have allowed the use of the non-linear joint elements for the vertical joints. The mesh details and the location of the expansion allowance materials are shown in Figure 17.19. The silica dome was constructed of three-dimensional solid isoparametric elements. As previously discussed, the expansion joints were constructed of three-dimensional interface elements. Because these elements are non-linear, an iterative procedure was used to converge on the correct expansion joint behavior (compression only). Since a tongue and groove was designed into the brick joint, it was assumed that shearing displacement normal to the spherical dome surface could not exist. That is, the mid-thickness nodes were coupled in the radial direction (motion normal to shell surfaces) to simulate the keying action of the tongue and

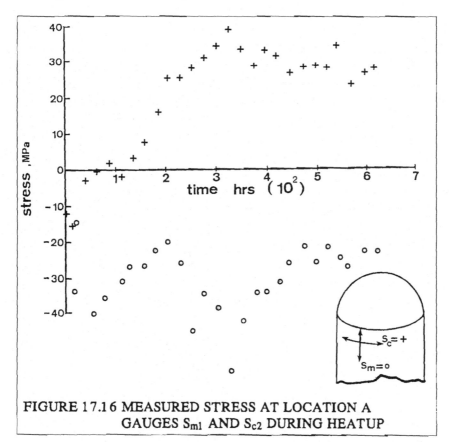

FIGURE 17.16 MEASURED STRESS AT LOCATION A
GAUGES S_{m1} AND S_{c2} DURING HEATUP

groove joints. Not shown with this model is the exterior vessel shell model. Only the cylindrical portion of the shell contributing to restraining the dome skews is shown. There are two equal lengths of cylindrical shell above and below the skew that provide the significant part of the restraint to the skew. Portions of the cylindrical shell can be quantified using classical cylindrical shell equations [8]. Assuming the normal force developed between the shell and skew is over a small finite distance, the assumption of a concentrated circumferential load is used. As shown in Figure 17.20, the maximum contributing length (L) of cylindrical shell on each side of the skew is defined as:

$$\beta L = 2.5 \qquad (17.6a)$$

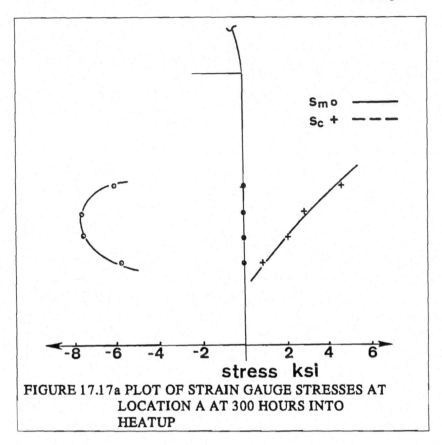

FIGURE 17.17a PLOT OF STRAIN GAUGE STRESSES AT
              LOCATION A AT 300 HOURS INTO
              HEATUP

or:

$$L = 2.5/\beta \qquad (17.6b)$$

where:

$$\beta = \sqrt[4]{\frac{3(1 - \nu^2)}{R^2 t^2}} \qquad (17.7)$$

where ν is Poisson's ratio (for steel, $\nu = 0.3$), R is the shell radius
and t is the shell thickness.

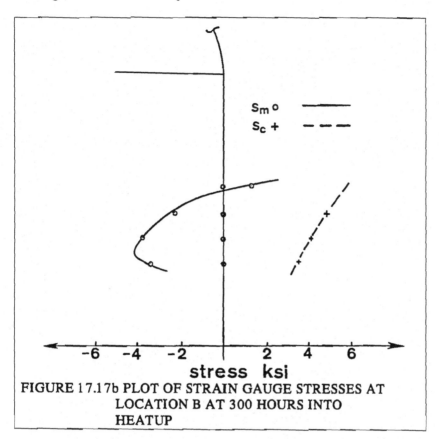

FIGURE 17.17b PLOT OF STRAIN GAUGE STRESSES AT
LOCATION B AT 300 HOURS INTO
HEATUP

Substituting the values of R and t from Figure 17.8 into Equation 17.7:

$$\beta = \sqrt[4]{\frac{3(1 - 0.3^2)}{(4877)^2(20)^2}} = 1/243$$

Substituting β into Equation 17.6a, and solving for L:

$$L = (2.5)(243)/1 = 608 \text{ mm}$$

Therefore, the effects of the skew horizontal reaction load is only

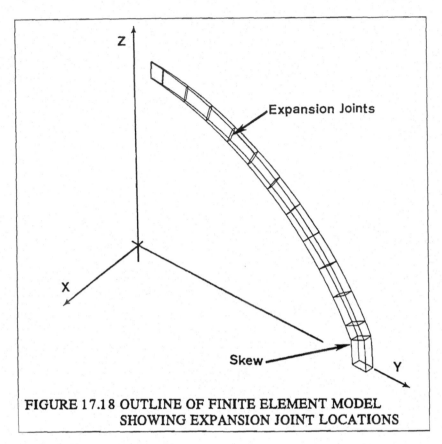

FIGURE 17.18 OUTLINE OF FINITE ELEMENT MODEL
SHOWING EXPANSION JOINT LOCATIONS

resisted by a short length (L) of the cylindrical vessel shell above and below the skew. The total length of the cylindrical vessel shell resisting the skew load is therefore, 1216 mm (2L). Beyond this distance the effect of shell moment, shell shear and shell displacement, resulting from the skew horizontal radial loading, decay to insignificant values. As shown in Figure 17.15, it was assumed that the center of the skew reaction forces would be at the center of the four sets of strain gauges, or at a distance of 750 mm measured vertically down from the weld connecting the cylindrical vessel shell to the spherical vessel shell. The strain gauge sets are spaced at equal intervals of 100 mm. Because of the 750-mm distance, the edge effects of the spherical shell will not influence

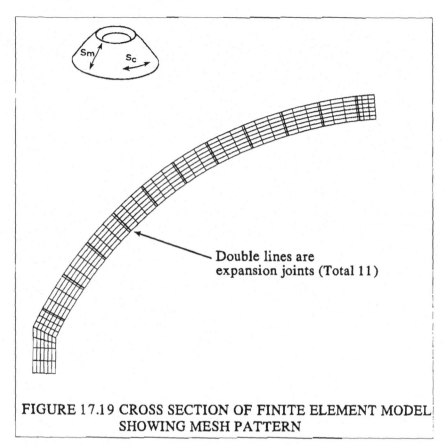

Double lines are
expansion joints (Total 11)

FIGURE 17.19 CROSS SECTION OF FINITE ELEMENT MODEL
SHOWING MESH PATTERN

the cylindrical shell behavior in the region of the skew.

Since strain gauge data were obtained on the shell stresses
during heatup, the analysis will include evaluation of the shell
stresses in the upper region of the stove shell wall. This information will assist in confirming gravity load and expansion forces
developed by the silica brick skew and silica brick dome. The shell
stress was influenced by three separate types of loads. The three
loadings are defined as:

1.   The horizontal (radial) thrust developed by the silica
        brick dome gravity load and expansion force.

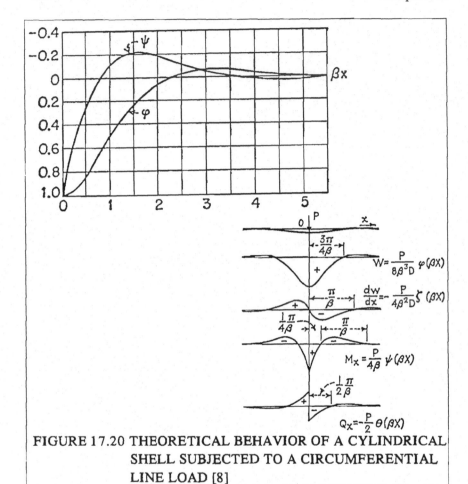

FIGURE 17.20 THEORETICAL BEHAVIOR OF A CYLINDRICAL
SHELL SUBJECTED TO A CIRCUMFERENTIAL
LINE LOAD [8]

2.   The horizontal (radial) expansion force developed by
       the skew and the adjacent region of the silica brick
       wall.

3.   The temperature gradient between the wall shell and
       dome shell.   This is not a through-thickness
       temperature gradient, but rather a membrane
       temperature gradient. A lower dome shell temperature
       would tend to pull the top of the shell wall inward. A
       higher dome shell temperature would tend to push the
       top of the shell wall outward.   Circumferential and

vertical stresses in the top of the shell wall would be developed compatible with the imposed dome displacements.

The predicted displacement of the silica brick dome, resulting from the thermal expansion of the brick and the restraint of the skew (resisted by the shell wall), is shown in the computer plot of Figure 17.21. Note that two primary circumferential hinges occur in the region of the skew. The circumferential hinge, at the skew, is on the coldface. The second circumferential hinge occurs in the second expansion joint, above the skew, and is on the hotface side. The second location is the approximate position in which pinch spalling occurred in the silica brick stove dome design (see Section III.B). Therefore, the solution appears to be valid, based on

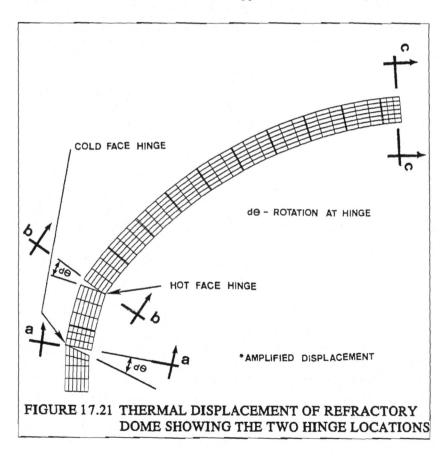

FIGURE 17.21 THERMAL DISPLACEMENT OF REFRACTORY DOME SHOWING THE TWO HINGE LOCATIONS

experiences with dome brick spalling. It should be mentioned that the hinge location could have been confirmed by evaluating the total strain energy versus hinge location. However, the nature of the finite element method will allow convergence to the correct hinge location.

The phenomenon of hinges is not unique to domes, and exists in other refractory shell structures, as discussed in the chapter on the sprung arch.

The highest circumferential stresses (about -11.05 MPa) occur near the skew, as shown in Figure 17.22. The section locations are described in Figure 17.21. Also shown in Figure 17.22 is the portion of the circumferential joint in bearing. The crosshatched box, above each through-thickness stress description, describes the bearing area of the joint. The existence of the hinges is primarily a result of the shell restraint on the skew and other effects previously discussed under theoretical considerations.

The meridional, or vertical, stress (about -0.86 MPa psi) is also highest near the skew, as described in Figure 17.22. The stresses in the upper regions of the dome are considerably less. As expected, the circumferential stresses are greater than the vertical stresses.

The second part of the analysis consists of the analysis of the upper stove shell using the radial skew expansion loads developed previously. These skew reaction loads were applied to the corresponding locations on the inside surface of the shell model. Figure 17.23 is a plot of the vertical stress (S_m) and circumferential stress (S_c) results on the outside surface of the stove shell. The skew reaction load was applied at the cold location of the skew. But, as previously defined, the center of the skew radial force was assumed to be applied at a location 600 mm down from the weld connecting the spherical portion of the vessel shell to the cylindrical portion of the vessel shell. As described in Figure 17.23, the vertical stress (S_m) on the outside surface reflects the outside surface component of the vessel bending stress. The predicted S_m stress behavior replicates the classically predicted behavior of the vessel shell bending moment (M_x) in Figure 17.20. The circumferential

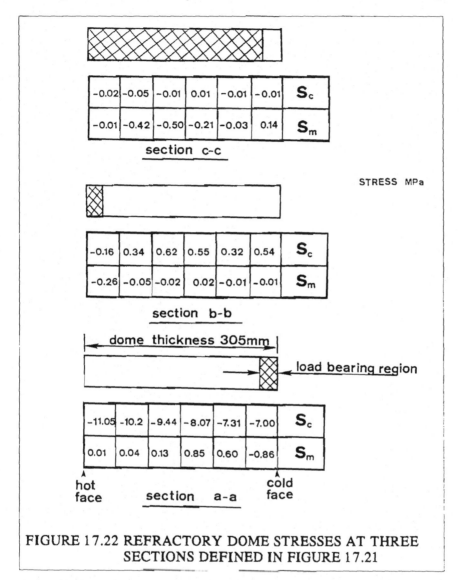

FIGURE 17.22 REFRACTORY DOME STRESSES AT THREE SECTIONS DEFINED IN FIGURE 17.21

stress (S_c) in Figure 17.23 also exhibits the same trend as the radial displacement W in Figure 17.20. The maximum outside stress at the skew is 196 MPa. The vertical stress oscillates and decays to values of -46 to -53 MPa on each side of the skew.

As noted in Figure 17.23, the gauges are assumed to be

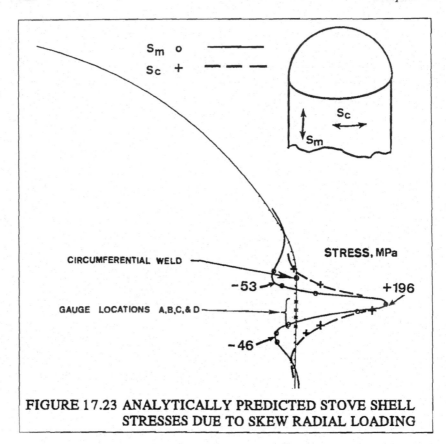

FIGURE 17.23 ANALYTICALLY PREDICTED STOVE SHELL STRESSES DUE TO SKEW RADIAL LOADING

below the position of skew. If the assumed skew load is raised to a higher position on the cylindrical shell model, the predicted stress contours agree with the measured stress values. A more detailed comparison is made in Figure 17.24.

The trends in the variations of vertical and circumferential stresses now compare quite favorably between the measured and analytical results. The circumferential stress (S_c) is tensile across the region of the gauges and increases when moving vertically upward toward the topmost gauges. The vertical stress (S_m) goes from negative (in the region of the lower gauges) to positive (in the region of the topmost gauges).

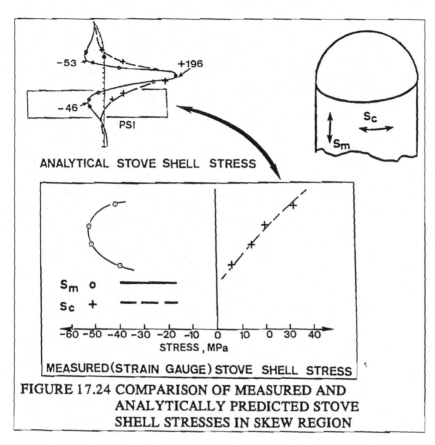

ANALYTICAL STOVE SHELL STRESS

MEASURED(STRAIN GAUGE) STOVE SHELL STRESS

FIGURE 17.24 COMPARISON OF MEASURED AND
ANALYTICALLY PREDICTED STOVE
SHELL STRESSES IN SKEW REGION

Since the measured shell stresses and the analytical shell stresses agree in both magnitude and general trends, stresses are considered to be an accurate evaluation of the actual stresses developed in the silica brick dome during heatup. These results also indicate that the installed position of the skew was apparently higher than anticipated and that the gauges were placed too low on the exterior surface of the shell. Perhaps the difference in vertical expansion of the interior refractory and the vessel shell caused the skew to grow upward more than expected. In either case, the outcome between the field test and analytical investigation was very encouraging and indicated that the predicted silica brick dome stresses were fairly accurate.

F. Summary of Spherical Silica Brick
Dome Investigation

As a result of this investigative analysis, a variety of summary items are provided which relate to the outcome of the investigation.

1. Using a 100% expansion allowance is not always the optimum approach in designing silica brick stove domes. During cooldown, a silica brick stove dome (and other portions of the stove brick work) with 100% expansion allowance may experience excessive brick movement resulting in necessary realignment before the next heatup.

2. In silica brick stove domes, two primary directions of thermal expansion demand consideration in the design of the expansion allowance. These directions are the circumferential direction and the meridional (vertical) direction. The circumferential expansion is directly influenced by expansion allowance material in the vertical joints. The meridional direction is directly influenced by expansion allowance material in the circumferential joints.

3. When the theoretical aspects of expansion allowance are considered, the vertical joints with expansion allowance material have a considerably greater contribution to expansion allowance than the horizontal joints with expansion allowance material. It should be noted that the beginning stages of silica brick stove dome design did not include expansion allowance material in the vertical joints.

4. The spalling locations, above the skew, encountered in another silica brick stove dome design are in agreement with the hotface hinge location (point of concentrated loading), as predicted by the brick dome model. Therefore, the effects of insufficient expansion

allowance, such as pinch spalling, will begin to occur initially at the hinge locations on the hotface side of the brick work. The hinges will occur at predictable locations of the dome as determined by principles of structural analysis for refractories. The study also indicates that abrupt changes in expansion allowance, especially in the lower region of the dome, are highly undesirable.

5. Based on the experience with percent expansion allowance in the spalled silica brick stove dome and the percent expansion allowance in the investigated brick stove dome design, the minimum allowance for thermal expansion in silica brick stove domes is about 60% for the silica brick with a 1.5% expansion at 1100 C. That is, the thickness of material which burns out during heatup should represent at least 60% of the total growth due to thermal expansion of a 1.5% expansion brick. This amount of expansion allowance is used in both the meridional and circumferential direction.

6. The field measurements showed that maximum radial expansion forces from the silica brick stove dome skew were developed after about 300 hours of heatup, or when the dome reached about 600°C.

7. The field measurements also showed that the skew evidently moved upward during heatup and, therefore, the skew radial expansion forces were above the location of the top strain gauges. Plots of the shell wall circumferential stresses and shell wall vertical stresses clearly show that the greatest expansion forces are developed by the skew brick and dome brick.

8. The analysis demonstrated that the thermal gradients in the shell wall and shell dome were not significant and that the skew radial expansion force developed by the thermal expansion of the silica brick stove dome was the predominant load in the shell wall.

9.   The analytically calculated silica brick dome radial skew loads, when applied to the shell, resulted in analytical shell stresses that agreed favorably with the measured shell stresses.   Therefore, the material properties used for the silica brick are valid for future designs of silica brick stove domes.

10.   The amount of expansion allowance used in the investigated silica brick stove dome is adequate for future silica brick stove dome designs.

In conclusion, this investigative analysis provides significant information for understanding the behavior of silica brick stove domes and basic engineering data for evaluating future silica brick stove dome designs.

IV. STEELMAKING LADLE REFRACTORY EXPANSION

A. Introduction

The objective of this chapter is to describe the necessary refractory mechanical material property data required when evaluating the thermal expansion behavior of refractory-lined cylindrical vessels and, more specifically, the compressive stress-strain data needed to define the MOE and other parameters associated with inelastic flow of the refractory.

The following investigative analysis examines and compares the measured expansion stresses of a refractory-lined cylindrical vessel (a steelmaking ladle) with the results obtained from a detailed finite element analysis.

B. Background

The following investigative analysis describes the influence of the stress-strain data used in characterizing the refractory lining of a cylindrical vessel. In our case, the field test results of an investigative study on a steelmaking ladle are used as the basis for the investigation (see Figure 17.25) [9].

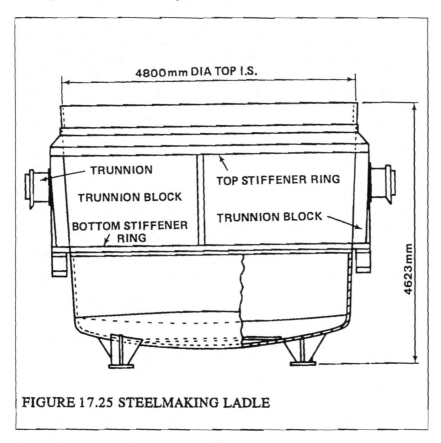

FIGURE 17.25 STEELMAKING LADLE

For those not familiar with steelmaking operations, the steelmaking ladle is used in the final stages of steelmaking. This ladle allows the transport of molten steel from the stationary steelmaking vessels to the continuous casters. Quite often additional steelmaking processing is conducted within the ladle. Therefore, the steelmaking ladle has become another process vessel in steelmaking.

In the past, the ladle was lined with fireclay brick. However, the recent emphasis on clean steelmaking practices has resulted in the use of cleaner, purer refractory linings. These refractories are also stronger and stiffer than the old fireclay linings. As a result of the use of stronger linings, such as high-alumina working linings, concern has been expressed over the structural integrity of the ladle shell due to the larger expansion loads imposed on the shell by these stronger linings. Appropriate refractory material properties are required in order to develop an accurate and dependable analytical prediction of ladle lining behavior. This study compares the analytical results with field test data. Updated analytical predictions are made using developments in refractory material testing which provide an improved definition of refractory structural material properties [10,11].

Since the primary interest in this study was in the cylindrical lining behavior, the ladle wall lining behavior was investigated. The ladle wall refractory lining design is described in Figure 17.26. As shown, the working lining is a 127-mm thick fired 70% alumina brick. The backup, or safety, lining consists of two parts: a 76-mm thick fireclay brick and a 51-mm thick alumina insulating brick.

C. Results of Field Test Measurements

The field strain gauge test was conducted on a newly lined ladle with the wall lining details as described in Figure 17.26. Strain gauges were placed on the exterior side of the wall vessel shell. Because of the limitation on the portable test measurement equipment, only the first heatup and process cycle could be recorded. As described in Figure 17.27, the cold ladle was initially subjected to a preheat. The maximum preheat temperature was about 1100°C. The ladle was then subjected to a series of *cooling* and *teeming* cycles. The cool portion of the cycle is the empty heated ladle going through various maintenance procedures prior to teeming. The heating or teeming portion of the ladle is the portion of the cycle when the ladle is filled with molten metal. During teeming, the ladle is serving as a molten metal transporting device. The molten metal temperature is at about 1630°. In our field test measurements, only the first cool and teem cycle was of interest. As noted in Figure 17.26, the initial preheat (or heatup) is fairly rapid. The field test thermocouple (TC No. 1) was placed at the back side or coldface of the working lining. The analytical results shown on the plot will be explained later.

The ladle vessel shell strain gauge result is shown in Figure

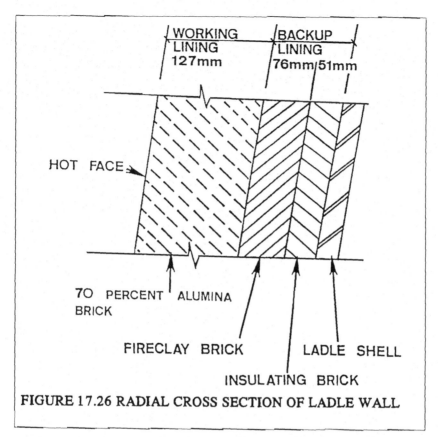

FIGURE 17.26 RADIAL CROSS SECTION OF LADLE WALL

17.28. This plot describes the variation in the ladle shell wall circumferential stress during the preheat and the first few cycles following the preheat. The maximum shell wall stress exists during the preheat, with a maximum value of about 100 MPa (16,000 psi), occurring after about two hours into the preheat. This stress remains at this level during the remainder of the preheat. The shell wall stress then begins to decrease in the subsequent cool and teem cycles. Of particular interest is the magnitude of the shell wall circumferential stress during preheat as compared to the magnitude during teeming. The maximum hotface temperature during preheat was about 1100°C. During teeming the molten metal temperature is approximately 1630°C. However, a proportional increase in expansion stress does not develop. As discussed in Chapter 6 on

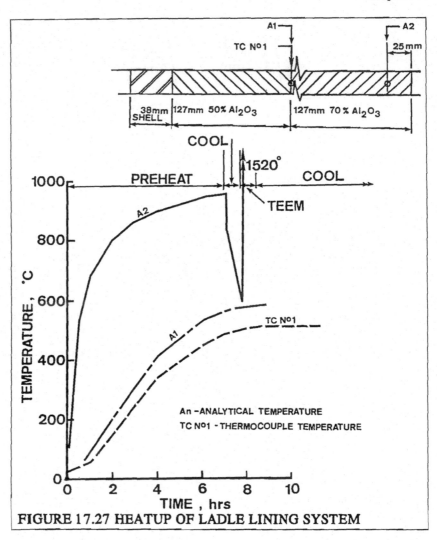

FIGURE 17.27 HEATUP OF LADLE LINING SYSTEM

compressive stress-strain behavior at high temperatures, the lesser shell wall stress is attributable to the inelastic behavior of the 70% alumina work lining at these higher temperatures. The lower shell wall stress during the short cooling event between the preheat and first teeming event is expected.

D. Results of Finite Element Analysis

The first portion of the finite element analysis was to deter-

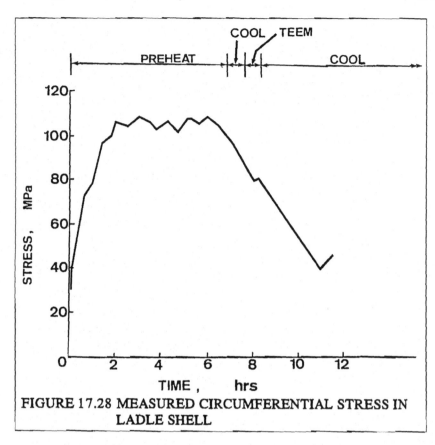

FIGURE 17.28 MEASURED CIRCUMFERENTIAL STRESS IN
LADLE SHELL

mine the lining and shell transient temperatures during the preheat
and subsequent process cycles. Since the shell wall measured
stresses are predominantly lining expansion stresses, the analyses
can be divided into two primary parts. Part I is the heat transfer
analysis used to determine the lining and shell transient
temperatures. Part II is the thermal stress analyses. The Part I
lining and shell temperatures are used to evaluate the lining
expansion stresses in Part II.

The thermal expansion stress is the predominant stress in the
ladle. The ferrostatic pressure develops a significantly lesser
circumferential stress. This stress is calculated as:

$$S_c = PR/t \qquad (17.8)$$

where R is the vessel shell radius, t is the vessel shell thickness and P is the ferrostatic pressure. The maximum ferrostatic pressure is estimated as:

$$P = D\delta \qquad (17.9)$$

where D is the full depth of the molten metal and δ is the liquid density of the molten metal. For the ladle defined in Figure 17.25 the maximum ferrostatic circumferential stress is:

$$
\begin{aligned}
S_c &= D\delta R/t \\
&= (3660)(2.30 \times 10^{-5})(4650)/(32) \\
&= (3660 \text{ mm})(2.30 \times 10^{-5} \text{ N/mm}^3)(4650 \text{ mm})/(32 \text{ mm}) \\
&= 12.21 \text{ MPa}
\end{aligned}
$$

The ferrostatic pressure stress in the vessel wall is shown to be small compared to the refractory lining expansion stress measured at levels of about 110 MPa. As a result, our efforts will concentrate on the lining expansion stresses and will assume that the field test stresses were predominantly lining expansion stresses.

The finite element model used in the analysis of the ladle wall is described in Figure 17.29. The model represents a unit through-thickness section of the ladle wall. Because of symmetry, only a quarter section of the through-thickness wall unit section was required. Figure 17.30 describes the boundary conditions for the non-linear compression-only stress condition of the mortar joints. Note that this model did not include the full dimensions of the brick lining components. In this model, the stress-strain behavior of the brick lining component in combination with the lining joint was based on a *smeared* effect. That is, the MOE and the inelastic effects were modified for the unit component section in the model to account for the effects of the mortar joint.

The results of the Part I transient thermal analysis are summarized in Figure 17.27. As shown, the analytically predicted transient temperatures of the lining coldface and the shell were consistently greater than the field test thermocouple measurements.

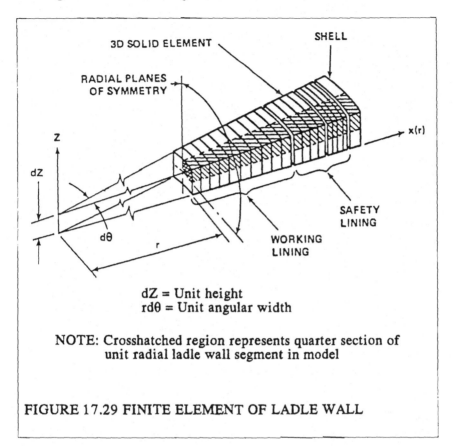

dZ = Unit height
rdθ = Unit angular width

NOTE: Crosshatched region represents quarter section of
unit radial ladle wall segment in model

FIGURE 17.29 FINITE ELEMENT OF LADLE WALL

This is attributable to the thermal material properties used in the transient thermal analysis. That is, the manufacturer's thermal material property data did not agree with the refractory lining material used in the lining. It was decided that since the analytical predicted temperatures of the lining and shell were both higher than the test results, the analytically predicted differential expansion between the lining and shell and resulting stresses would not be significantly different from the field measured results.

The Part II thermal expansion stress analysis was conducted using three different assumptions regarding the definition of the mechanical material properties of the 70% alumina brick working lining. As will be described, the backup lining plays an insignifi-

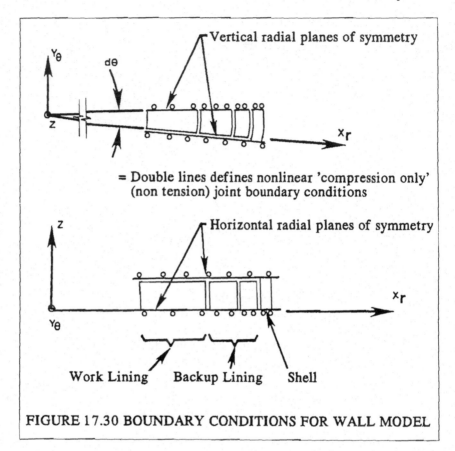

= Double lines defines nonlinear 'compression only'
(non tension) joint boundary conditions

Work Lining   Backup Lining   Shell

FIGURE 17.30 BOUNDARY CONDITIONS FOR WALL MODEL

cant role in developing expansion forces against the shell.
Therefore, the mechanical material properties of the backup lining
were less important in evaluating the thermomechanical behavior of
the ladle wall.

The first thermal expansion stress analysis was conducted
using available ultrasonic modulus data for Manufacturer A's 70%
alumina brick.   Figure 17.31 and Table IV describe the ultrasonic
elastic modulus data as a function of temperature.   As shown, the
elastic modulus remains constant over the range of operating
temperatures imposed on the ladle lining using the Part I calculated
lining and shell transient temperatures, the calculated lining
expansion stress in the shell wall is described in Figure 17.32.

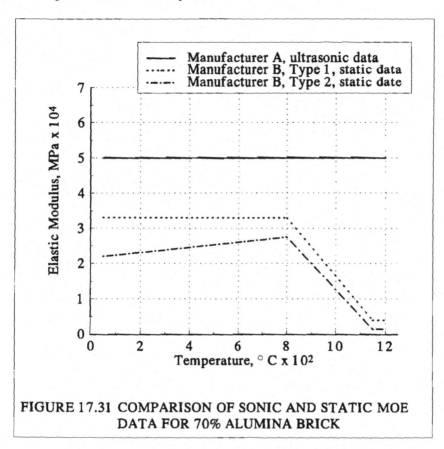

FIGURE 17.31 COMPARISON OF SONIC AND STATIC MOE
DATA FOR 70% ALUMINA BRICK

As described, the calculated shell wall circumferential stress is nearly three times greater than the measured shell stresses during preheat. During the teem portion of cycle, the calculated shell wall stress is about four times greater than the measured shell stresses.

It is obvious that a severe discrepancy exists between the calculated and measured shell stresses. Using ultrasonic elastic modulus data, the lining develops an unrealistically high expansion force. It can also be concluded that the ultrasonic elastic modulus does not represent the true elastic modulus of the refractory lining material. In addition, the ultrasonic test data do not provide any information with regard to the inelastic behavior at the higher temperatures.

Table IV

Comparison of Elastic Modulus
for 70% Aumina Brick

Temperature °C	Manufacturer A Ultasonic Modulus x 10^4 MP	Manufacturer B * Static Modulus x 10^4 MPa	
		Type 1	Type 2
20	4.8	3.17	2.14
800	4.8	3.10	2.62
1100	4.8	1.10	1.52
1200	4.8	0.28	0.19
1300	4.8	---	---

*Values from Reference 10.

A second stress analysis was conducted using creep data for 70% alumina brick. In this second analysis, the secondary creep equation was used in which the secondary creep strain (ϵ) was defined as:

$$\epsilon_c = Af(s)t^m\sigma^h e^{-\Delta Hc/PT}$$

where Af(s) is equal to 2.32 x 10^9, m is equal to 1.0, h is equal to 1.0 and -ΔHc/P is equal to -5.82 x 10^4. Temperature (T) is in degrees Kelvin and time (t) is in hours. It should be noted that the creep data were developed for a compressive stress of about 0.17 MPa (25 psi). In the second analysis, the ultrasonic elastic modulus was used as in the first analysis. This analysis using the ultrasonic elastic modulus and the creep material property data, resulted in no significant difference in the calculated stresses from the first analysis at temperatures below about 1200°C. Therefore, it was concluded that the secondary creep data were not satisfactory. The second analysis was then modified using a reduced coefficient of thermal expansion. Typically, 70% alumina brick has a coefficient of thermal expansion of about 6.3 x 10^{-6} mm/mm°C (3.5 x 10^{-6} in./in.°F). For this modified analysis, the coefficient was reduced to a value of about 2.1 x 10^{-6} mm/mm°C. The results of the

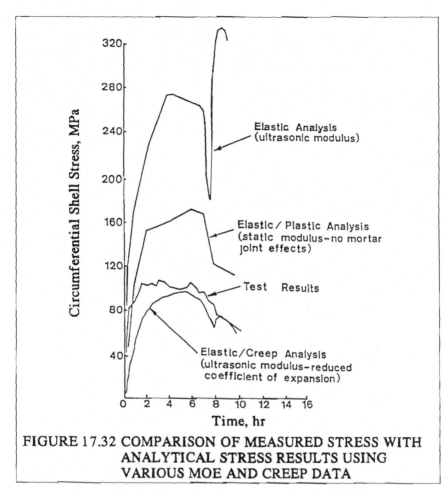

FIGURE 17.32 COMPARISON OF MEASURED STRESS WITH
ANALYTICAL STRESS RESULTS USING
VARIOUS MOE AND CREEP DATA

analysis are plotted in Figure 17.32. Using a reduced coefficient of
thermal expansion in combination with the ultrasonic elastic
modulus and the defined secondary creep, the analytically calculated
shell stress results closely resemble the field-measured results.
These results assist in confirming that the ultrasonic modulus does
not represent the true elastic modulus of the refractory working
lining, and also that the creep data do not represent the true creep
response of the working lining. One should not have to resort to
such a modification or fudge factor to produce reasonable results.

The resulting circumferential expansion stresses in the vessel shell from the second modified analysis are described in Figure 17.33. As shown, the working lining develops nearly all of the expansion force and the backup lining contributes an insignificant expansion force. Negligible inelastic strain is developed during the preheat (temperatures less than 1100°C). However, during teeming (temperatures between 1100 and 1630°C), compressive creep

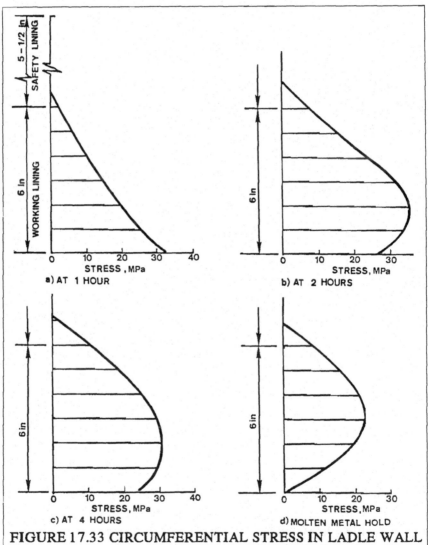

FIGURE 17.33 CIRCUMFERENTIAL STRESS IN LADLE WALL
LINING DURING VARIOUS HEATUP TIMES

straining did occur but was not sufficient in providing a lining expansion response that resembles the field measured data. As a result, the adequacy of the compressive creep data (developed at laboratory compressive stresses of 25 psi) was questionable since the average compressive lining stresses in the tested ladles were estimated to be about 34 MPa (5000 psi).

The third analysis used static compressive stress-strain data of a 70% alumina brick to define the elastic modulus. The homogeneous static compressive stress-strain data for a 70% alumina refractory brick are described in Figure 17.34 [10]. The methodology of converting the compressive static stress-strain data to a static elastic modulus (E_{sc}) is illustrated in Figure 17.35. A bilinear stress-strain curve was used in the structural program to

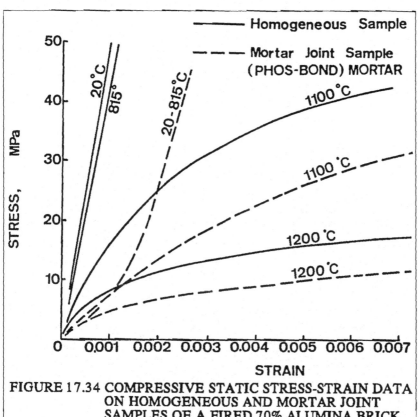

FIGURE 17.34 COMPRESSIVE STATIC STRESS-STRAIN DATA ON HOMOGENEOUS AND MORTAR JOINT SAMPLES OF A FIRED 70% ALUMINA BRICK

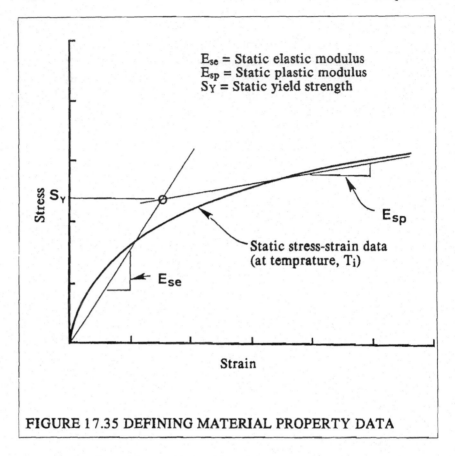

FIGURE 17.35 DEFINING MATERIAL PROPERTY DATA

model the elastic/plastic response of the material. The onset of yielding is defined by the yield point stress (S_Y). The slope of the stress-strain response beyond the yield point is defined as the plastic modulus (E_{sp}). The elastic modulus was calculated using the slope of each curve defined by a 14-MPa compressive stress and the corresponding strain. Table IV provides a comparison of the elastic modulus (ultrasonic modulus) used for the 70% alumina brick (Manufacturer A) in the initial ladle study [8] and the elastic modulus (static stress-strain) for two similar 70% alumina brick materials (Manufacturer B) [10,11]. A plot of the Table IV data is described by Figure 17.31. The static elastic modulus data differ from the ultrasonic elastic modulus data in that (1) the static modulus results are less than the ultrasonic modulus results by a significant margin and (2) the static modulus data reflect the temperature-dependent inelastic effects of the 70% alumina brick.

The material data of Manufacturer B's alumina brick was chosen for the analysis since this 70% alumina brick was most similar in all aspects to the Manufacturer A's 70% alumina brick used in the ladle tests. Table V summarizes the resulting complete elastic-plastic bilinear stress-strain data used to represent the 70% alumina working lining.

The results of this analysis are summarized by comparing the calculated circumferential shell stress with the field test results. As shown in Figure 17.32, the general shape of the third analysis stress results closely resembles the test results. However, the analysis stress results were about 50% greater than the test results. The analysis was conducted using static compressive stress-strain data for homogeneous 70% alumina brick samples. More recent stress-strain data on 70% alumina brick samples with mortar joints (see mortar joint compressive stress-strain data, Figure 17.34) show that mortar joints tend to cause an additional softening of the refractory lining. The stress-strain data with mortar joints show about a 50% softening due to the presence of the mortar joints. The mortar joint test samples had a 1.5 mm mortar joint for a sample length of 50 mm. It can be concluded that the inclusion of the mortar joint effects would result in an analytical solution that more closely resembles the field test results.

E. Summary

The results of the analyses show that ultrasonic elastic modulus data do not provide a true representation of the stress-strain relationship for refractory materials. The purpose of this study is not to diminish the usefulness of ultrasonic modulus data. For the refractory engineer, the ultrasonic modulus assists in ranking the refractoriness of the material; however, for the structural engineer, the static stress-strain data provide a more realistic

Table V

Mechanical Material Data for
70% Alumina Brick

Temperature °C	Elastic Modulus* x 10⁴ MPa	Plastic Modulus* x 10⁴ MPa	Yield Stress MPa	Poisson's Ratio
20	2.41	1.93	48.3	0.07
800	2.41	1.38	48.3	0.09
1100	1.45	0.80	14.5	0.17
1200	1.03	0.08	7.8	0.14
1300*	0.34	0.01	3.45	0.14
1650*	0.0007	0.00003	3.45	0.14

*Data extrapolated beyond the maximum test temperature of 1200°C.

estimation of elastic/plastic mechanical properties of refractory materials for structural analysis.

The reason for the differences in the analyses when using the ultrasonic data is best described in Figure 3.4. The ultrasonic modulus is defined by taking advantage of the high-frequency vibration characteristics of the sample. However, this results in the evaluation of an elastic modulus at a very low stress state. As shown in Figure 17.36, the tangent to the actual stress-strain curve at a low stress state results in a high elastic modulus. With the actual stress-strain curve defined, the method of analysis (using the elastic or elastic-plastic mechanical property definition) will provide improved accuracy of the predicted stress. Since thermal expansion is a strain-controlled load, the elastic/plastic solution, which traces the actual stress-strain curve, will provide the best solution.

F. Conclusions

The results of the analyses using static stress-strain data for

70% alumina refractory brick are very encouraging and provide considerable confidence in using analytical methods in combination with the associated bilinear elastic/plastic mechanical material property definition in predicting the thermomechanical behavior of refractory structures. The following conclusions are made as a result of this study [12]:

1. The temperature-dependent static stress-strain material property data provide a comprehensive method of defining refractory material mechanical behavior.

2. The elastic-plastic definition of a refractory lining material obtained from the static stress-strain material property data provides an improved method of predicting the thermomechanical behavior of the refractory structure.

3. As described in this study, the appropriate analytical methods and the use of static compressive stress-strain data are applicable for similar refractory-lined vessel and furnace investigations.

4. The incorporation of mortar joint material with the static compressive stress-strain data is needed to provide a realistic definition of the elastic-plastic definition of the refractory lining material.

5. Refractory creep data must be defined for more realistic values of compressive stress.

REFERENCES

1. Schacht, C. A, Stress-Strain and Other Thermomechanical Environments Imposed on Refractory Systems in Steelmaking Operations, American Ceramic Society, Ceramic Trans., Advances in Refractory Technology, Vol. 4, pp. 205-242 (1990).

2. Harbison-Walker Handbook of Refractory Practice, First Edition, Harbison-Walker Refractories, One Gateway Center,

Pittsburgh, PA, 1992.

3. Schacht, C. A., A Viscoelastic Analysis of A Refractory Structure, Ph.D. Thesis, Carnegie-Mellon University, April, 1972.

4. Schacht, C. A. and Hribar, J. A., An Elastic Analysis of A Refractory Sprung Arch, American Ceramic Society, Bulletin, Vol. 53, No. 7, pp. 528-531 (1974).

5. The Making, Shaping and Treating of Steel, Ninth Edition, United States Steel, Pittsburgh, PA, 1971.

6. Palin, F. T., Stress Calculations for The Sections of No. 2A Blast Furnace Stove at Altos, Hornos De Vizcaya, Spain, Constructed with Silica Brick, British Ceramic Research Association, September 9, 1984.

7. ANSYS, Engineering Analysis System; Swanson Analysis Systems, Inc., P.O. Box 65, Houston, PA 15342.

8. Timoshenko, S. and Woinowsky-Krieger, S., Theory of Plates and Shells, Second Edition, McGraw-Hill Book Company, Inc., 1959, pp. 533-558.

9. Schacht, C. A. and Abarotin, E. V., Structural Behavior of Teeming Ladles Exposed to Clean-Steel Operations, Iron and Steel Engineer, Vol. 61, No. 8, pp. 33-40 (1984).

10. Stett, M. A., Measurement of Properties for Use with Finite Element Analysis Modeling, Ceramic Engineering and Science Proceedings, American Ceramic Society, Columbus, OH, Jan.-Feb. 1986.

11. Alder, W. R. and Masaryh, J. S., Compressive Stress-Strain Measurement of Monolithic Refractories at Elevated Temperatures, American Ceramic Society, Advances in Ceramics, Vol. 13, Columbus, Ohio (1985).

12. Schacht, C. A., Improved Mechanical Material Property

Definition for Predicting the Thermomechanical Behavior of Refractory Linings of Teeming Ladles, Journal of American Ceramic Society (Vol. 76), No.1, pp. 202-206 (1993).

Index